IN QUEST OF QUIET

MEETING THE MENACE
OF NOISE POLLUTION

call to citizen action by
HENRY STILL

fɾeð keɾneɾ/publishing pɾojects
in association with
STACKPOLE BOOKS

IN QUEST OF QUIET

IN QUEST OF QUIET

Copyright © 1970 by
HENRY STILL

Published by
STACKPOLE BOOKS
Cameron and Kelker Streets
Harrisburg, Pa. 17105

ISBN 0-8117-0891-8
Library of Congress Catalog Card Number 77-123404
Printed in U.S.A.

Contents

Peace and quiet crumble before the collective din of dishwasher, vent fan, food blender, hi-fi, television, quarreling children, and flimsy walls which bring the neighbor's lawn mower into the living room.

Section III The Case for Tranquility

Section IV The Nature of Sound

Preface

NOISE IN THE areas of America where most people live is twice as loud today as it was fifteen years ago. This attack upon our physical and nervous systems may double again in the next ten to fifteen years unless we pay careful and concerted attention to techniques for muffling the machinery of our industrial civilization.

Based upon a large number of technical and scientific studies, this book is intended to fill two functions: (1) identify the main sources in the rising tide of sound which threatens to engulf us and (2) suggest some ways in which we may move forward to a quieter world.

The table in Chapter 11 may be used as a guide to the average reader on the measurement of sound by decibel level in fluctuations of atmospheric pressure. The decibel is a measuring unit named for Alexander Graham Bell, who is better known for the invention of the telephone than for his excellent early studies in the nature of sound. An important point to remember in relation to the scale in Chapter 11 is that an increase of six decibels is equal to a doubling in loudness. Likewise, a six-decibel decrease results in a sound roughly half as loud.

Sound frequency, or pitch, will be designated in Hertz (abbreviated Hz), which denotes cycles per second. The frequency range in which sound is audible to most people extends from about 30 Hz to 18,000 Hz.

The reader who desires a more comprehensive discussion of sound and its effects, both physical and psychological, upon human beings is invited to refer to the final section of the book titled "The Nature of Sound."

Introduction

IN A SUBURBAN home in the Midwest, a father physically attacked his teen-age son because the boy refused to turn down the volume of rock and roll music booming from the hi-fi set.

In New York City, a young woman hammered an automobile with her high-heeled shoe because she had been driven to distraction night after night by the cacophony of automobile horns in the street below her twelfth-floor apartment.

On a metropolitan freeway, normal workday traffic ground to a rumbling, congested halt. While the multiple lanes of autos were stopped, two drivers leaped out of their vehicles and engaged in a battering fist fight in a misdirected effort to attain relief from the tension of frustration and noise.

Residents living near Los Angeles International Airport filed lawsuits claiming billions of dollars in damages for physical and mental harm caused by the incessant high noise level from jet planes.

In the Bronx, a man with a nighttime job shot and killed one of several boys whose noise at play disturbed his daytime sleep.

These apparently isolated cases of Americana are not exceptional for their flashes of bitter frustration or temporary insanity. Rather they indicate how millions of people everywhere are suffering emotional and physical illness from the rising tide of noise which assaults us from every direction at all hours of the day and night. Seldom anymore are we able to escape.

Although it is possible for the human system to adapt to a multitude of adverse conditions, it is becoming more evident each year that man, in his headlong rush to greater technological achievement, is polluting his land, water, and air—his entire environment—with substances which threaten all living things, himself included.

In the United States and most other industrialized nations of the world, virtually all freshwater streams and lakes are polluted with human sewage, chemical wastes from industry, and salts from agricultural land. The land itself is littered with trash, glass, and metal, including at least 7 million automobiles junked each year. The atmosphere becomes more murky each year with the smoke and fumes emitted by industrial and power plants and the millions of automobiles which exude carbon monoxide, evaporated gasoline, nitrogen oxides, and other harmful chemicals.

Despite piecemeal efforts to solve some of these problems, it was not until 1970 that President Richard Nixon presented a unified program, which if persistently followed, will restore some of the human graces which have been gradually lost through more than a century of industrial and commercial growth. Noise is as much a part of environmental pollution as smog or garbage. In the areas of America where most people live, noise on the average, both day and night, is twice as loud today as it was fifteen years ago. It threatens to be twice again as loud by the end of the 1970s unless we pay careful and concerted attention to techniques for muffling the machinery of our industrial civilization.

We have long considered noise an annoyance. Now it has become a danger at levels which cause hearing loss, mental sickness, and physical harm. It may well be a slow agent of death contributing to hypertension and high blood pressure, a major killer in modern society. The human ear, designed by nature originally to help primitive men survive in a dangerous environment, is ever alert. It cannot close itself (like an eyelid) against unwanted sound and therefore makes itself and the entire nervous system vulnerable to harm from noise of a thousand sources.

Noise existed even in ancient times. We can imagine the screech of stone and the groans and death cries of the slaves who built the great pyramids of Egypt. Camels, donkeys, and the hawkers of silk and swords created a din where merchant trails crossed in the old cities of Europe and Asia. And above all, the crashing noise of battle disturbed the natural silence wherever armies followed the bloody tracery of conquest through history.

Through all eras, however, the only lasting definition of noise has been *unwanted sound*, because even the sound of battle is music to certain ears. In general, the concept of noise follows men at work, particularly where men have harnessed animals or machines to do their work for them. Noise grows to its most irritating loudness where men and machines congregate in greatest numbers to perform life-sustaining work for each other.

The iron machine, born with the Industrial Revolution nearly two centuries ago, accounts for the vast bulk of aural pollution which degrades our environment and threatens in the future to produce a population partially deaf and in a perpetual state of mental agitation. All progress since invention of the steam engine may well be measured in the noise levels created by the mechanical monsters which click, whir, thump, roar, and shriek while multiplying the energy available in the muscles of a man's hands and feet. Since 1782, the insatiable greed for growth and power has been heard in the grinding of coal and oil from the earth's bowels, the screech of iron ore, the din of the boiler factory, and the whistle of steam locomotives moving people and commodities to the far coves of the continents.

We have lived so long with the clatter of automobile, tractor, and airplane that we forget the noisy cities of not so long ago. An anonymous contributor of the *Scientific American* vividly portrays the sound of London in the year 1890 when the horse was yet king.

". . . And after the mud," he wrote, "the noise surged like a mighty heart-beat in the central districts of London's life. It was a thing beyond all imaginings. The streets of workaday London were uniformly paved in 'granite' sets . . . and the hammering of a multitude of iron-shod hairy heels, the deafening side-drum tattoo of tyred wheels jarring from the apex of one set [of cobblestones] to the next, like sticks dragging along a fence; the creaking and groaning and chirping and rattling of vehicles, light and heavy, thus maltreated; the jangling of chain harness, augmented by the shrieking and bellowings called for from those of God's creatures who desired to impart information or proffer a request vocally—raised a din that is beyond conception. It was not any such paltry thing as noise. It was an immensity of sound."

The immensity of sound which plagues us today threatens to leave us no place to hide tomorrow. We must recognize that noise is a direct product of people in motion. The population of America and the world is growing at an alarming rate and environmental pollution—of which noise is one element—grows along with it. Millions of machines are at work building homes, factories, skyscrapers, and paving streets just in time for other machines to come along and rip them up again for utility installation or repair. Nearly 100 million autos, trucks, and buses burn that many gallons of fuel every half-hour in the noisy internal combustion engine. Roaring coal and oil furnaces pour megawatts of power from steam-generating plants. The airplane, once a toy for millionaires and fools, has become a prime mover of people and freight, piercing the ears of millions with the shriek of jet engines. As we build supersonic wings, the boom of our passage strikes a blow to the earth below, and the electronic arts have given our teen-agers the power to attack our ears with amplified noise of unbearable levels.

There was a time when, if we did not enjoy the noise and dirt of the railroad or highway, we could move away from it. That's how suburbia was born, the migration away from the noise and ugliness of heart cities. Indeed, increasing noise may well be an elemental cause of urban riots among the underprivileged who are unable to move away beyond earshot of a neighbor's quarrels and gurgling toilet.

For everyone, however, it becomes more difficult with each passing year to find a quiet place. Automobiles, barking dogs, bulldozers, and crying children are everywhere because as our numbers increase, the buffer space between people grows less.

There is the temptation to oversimplify the situation by stating that noise is the inevitable companion of man's efforts to serve the needs of an exploding population. The noise we suffer is indeed such a companion, but the companionship is not inevitable. The fact that we can now make fairly accurate statistical projections of population trends is the first hopeful sign that we can use those projections as a basis for techniques by which more of us can live closer together in greater harmony in the future.

Such techniques, without question, will require great effort. As our population grows, the movement of people into metropolitan areas is growing even faster despite the fact that most cities today are generally ugly, noisy places in which to live. At the beginning of 1970, it was estimated that 140 million of the 205 million U.S. residents were city dwellers. If present trends continue, the country's population will increase to 300 million by the turn of the new century with 80 percent living in cities. The poet Shelley wrote in the early 1800s that "Hell is a city much like London, a populous and smoky city." In 1969 Lord Ritchie-Calder, British newspaperman and author, predicted that by the year 2025 all of the people in the world will live in cities, the largest of which would be 186 times larger than greater London today. If these projections were to hold true, along with comparable increases in today's noise levels, the world of 2025 might well be populated by a race of half-deaf neurotics. In fact, if noise and other environmental pollution were to continue unabated, the people living then would be half-mad savages driven to maniacal destruction in order to preserve a small area of cleanliness and quiet around them.

Noise, according to Dr. Norman K. Sanders of the University of California at Santa Barbara, is becoming a major factor in degrading the quality of our environment. William H. Stewart, former U.S. surgeon general, stated: "Calling noise a nuisance is like calling smog an inconvenience. Noise must be considered a hazard to the health of people everywhere."

The most direct hazard posed by noise is the destruction of hearing ability. For more than a century scientists have been documenting the loss of hearing due to noise among workers in heavy industry. In 1968, the Federal Council of Science and Technology estimated that from 6 million to 16 million men and women in America today are working under noise conditions which will damage their hearing. Outside industrial factories, however, the noise of industrial civilization has risen so high, particularly in major cities, that more than one-half of our entire population—more than 100 million people—may experience a gradual and partial deafness.

As noise levels rise, voice communications become more difficult. This condition often is responsible for injury and death in hazardous occupations.

A number of scientific studies now show that rock and roll music, amplified to ear-splitting levels in enclosed dance halls, is reducing many teen-age ears to the hearing dullness of a 60-year-old man.

Millions of people around the nation's major jetports are subjected to such high-level and high-frequency sound that many are driven to psychological and even physical illness, caused primarily by their inability to control the noise.

In American homes normal disagreements and friction among family members is intensified by levels of noise from kitchen appliances and other labor-saving machines which may well be contributing to the unrest of young people and decay of the family, the basic unit of society.

Medical research is beginning to show that loss of hearing and mental distress are by no means the only ill effects of noise. Loud sounds cause blood vessels to constrict, the skin to pale, muscles to tense, and adrenal hormone to be injected into the bloodstream. Loud noise can increase body tensions, which then can affect the blood pressure and functions of the heart and nervous system. Because of the ear's inability to shut out unwanted sound, these dangerous physiological changes occur under the impulse of noise whether a person is asleep or awake.

In a word, noise may well be hastening the day of death for millions of people in all the technologically developed nations of the world.

It is now past time for the scientists and engineers—and especially the politicians—of the world to take note of this serious environmental degradation and start turning our world back toward quiet and tranquility.

SECTION I

THE SOUNDS OF
CIVILIZATION

CHAPTER 1

Peace and quiet crumble before the collective din of dishwasher, vent fan, food blender, hi-fi, television, quarreling children, and flimsy walls which bring the neighbor's lawn mower into the living room.

A
Man's
Castle . . .

"TO OUTTALK EVE, Adam probably discovered shouting."
That thought, although perhaps not original, was expressed recently by J. C. Webster, a specialist on speech intelligibility at
the Naval Electronics Laboratory Center in San Diego, California.
In a sense, he summed up the thoughtful man's need for privacy and quiet, particularly in his own home. Schopenhauer, in
the eighteenth century, already had experienced the shock of
noise when it interrupts mental concentration and jars a sensitive person out of serenity. He complained bitterly about the

number of persons who use their "vital powers" in the form of "knocking, hammering, and tumbling things about. . . . Certainly there are people, nay very many, who will smile at this, because they are not sensitive to noise. It is precisely these people, however, who are not sensitive to argument, thought, poetry or art, or any kind of intellectual impression, a fact to be assigned to the coarse quality and strong texture of their brain tissues."

"Noisy interruptions prevent concentration," the philosopher wrote. "This is why the most eminent intellects have always been most strongly adverse to any kind of disturbance, interruption and distraction, and above everything, that violent interruption which is called noise. Noise is the most impertinent of all interruptions, for it not only interrupts our own thoughts but disperses them."

Following Schopenhauer, Oscar Wilde in 1882 wrote his "Impressions of America," including the complaint that "America is the noisiest country that ever existed. One is waked up in the morning not by the singing of the nightingale, but by the steel worker. It is surprising that the sound practical sense of the Americans does not reduce this intolerable noise. All art is based on exclusive and delicate sensibilities, and such continual turmoil must ultimately be destructive to the musical faculties."

Wilde could not have foreseen that in "the noisiest country that ever existed" modern-day music itself would be destructive to the musical faculties, or that many other faculties could be thrown out of balance by the rising clash of noise, including the ability to make lucid decisions upon important matters. Probably the one enduring quotation from President Richard Nixon's inaugural address in 1969 is his statement that "until we can stop shouting at one another, we cannot hear one another."

. . . The Home

The places where we need most to hear one another, and need isolation from all distracting sound, are the places where we live most of our lives—in the office, schools, hotel rooms, and most of all, our homes. It is in the home that we kick off our shoes and, in "peace and quiet," marshal our physical and mental forces for the next day's encounter with the freeway, the commuter train, the construction clatter on downtown streets, and perhaps the noise level of an industrial job. Home is the place where complete and restful sleep is absolutely vital to the recuperation and partial rejuvenation from yesterday's jangling experience.

Home, however, is no longer a quiet place for those majority millions of Americans who live in the cities and suburbia. Aside from family quarrels and the restless feet of youngsters running through a hallway (both of which haven't changed much in this century), at least 90 percent of our indoor noise is the result of modern conveniences with which we have surrounded ourselves for easier living and recreation.

Beginning in the kitchen, where mother is beginning to prepare dinner, the refrigerator (which has been well silenced through 40 years of engineering) offers the first whisper of perhaps 10 decibels. The sound of frying meat adds another 10. A food blender jumps the level another 20 dB, the automatic clothes washer 20 more. The automatic dishwasher, set into the kitchen counter for convenience, passes its vibration into the wood floor and walls, jumping the noise level another 30 dB. The hot air furnace fan (or air conditioner) adds another 10 to 20 dB to the overall, and here, in the quiet of a modern suburban home, the total noise level has reached 90 to 100 dB, which has been judged to be beyond the danger point for hearing loss in industrial situations.

Now add one more ingredient. The teen-age son of the house puts a rock and roll record on the living room stereo set and suddenly the sound level in that house may be 120 to 140 dB, the level associated with actual auditory pain. That combination of sound, though perhaps differing in frequency, may be equal to the noise experienced by a person standing 100 feet from a jet airliner at takeoff. Granted that such a total combination of noise is not present in the average home more than two hours per day, and thus may not constitute a threat of permanent hearing loss, the fact remains that a person should never be exposed to such noise levels. Adding the indoor noise to the impingement upon our walls by the sound of traffic, the neighbor's barking dog and power lawn mower, and a sonic boom or two, we reach a condition most of our waking hours in which it is impossible to find a quiet room in the house.

Internal noise need not approach these harmful levels to cause intense annoyance and nervous tension, particularly among women, who often are more sensitive than the male of the species. A weary mother may retire early but be unable to sleep because the music from a son's radio sounds through the adjoining wall. When this ends, she will be aroused again by another son, returning from a date, walking across a creaking wood floor. She must wait, then, until he has finished elaborate ablutions for bed (with running water clearly audible through the underfloor pipe transmission system). By now it may be midnight or later and mother's frustration level has peaked, requiring another hour for her to relax into sleep after the noise has stopped. Often, to forestall the multitude of night noises, she will take recourse to sleeping pills, or turn on an electric fan or air conditioner to provide a steady level of sound to mask the intruding noise.

Although we have considered the sound level in an average, lower middle-class home, the exacerbations of noise are even more disturbing to apartment dwellers. In a city apartment, a family will hear not only its own sound but will be distracted by the sound of a party in the apartment below, footsteps on the ceiling above, a family quarrel through the bedroom wall, and in the deep of night, the gurgle of plumbing. This is annoying not only to the listener, but causes (in poorly constructed apartment houses) a constant level of apprehension that all of the sounds a person makes are equally audible to the neighbors. No psychologist or sociologist has been able to reduce a set of reliable statistics measuring this lack of acoustic privacy in terms of happiness or unhappiness, but many investigators hold that this factor—the inability to escape to a quiet, private place—is one of the most serious causes of riots among the underprivileged dwellers of our major cities.

Based upon numerous studies, the Public Health Service in 1967 issued a set of recommended criteria for noise levels which should be permitted in various rooms. These range from 25 to 30 dB in broadcast studios up to 60 dB in sports coliseums. Relating to homes, the limits specify a level of 35 to 45 dB in a bedroom with an additional 5 to 10 dB permitted in living and recreation rooms. (It should be noted that room air conditioners manufactured before 1957 commonly produce levels of 40 to 55 dB in sleeping areas.)

Needless to say, these ideals are seldom achieved because of noises originating both inside and outside of the home. When determining annoyance factors, it is additionally interesting to note that most families do not object to *their own* noise (with the exception of friction between individual family members engaged in different activities) but complain of noise caused by others. A British survey of people living in a 36-square-mile section of London, showed that most residents are annoyed most by noise when they are at home (rather than outdoors or at work). The things they complained about included: road traffic, 36 percent; aircraft, 9 percent; trains, 5 percent; industry and construction, 7 percent; domestic appliances, 4 percent; neighbors' impact noise, 6 percent; children, 9 percent; adult voices, 10 percent; radio and TV, 7 percent; and pets, 3 percent.

One of the major reasons why noise in the home has become a serious and growing problem, although it seems paradoxical in this age of advanced affluence and technology, is that homes (including apartments) are not built as well as they used to be. This comes as no surprise to anyone who can remember the thick walls, heavy doors, and solid floors of a bygone age, but it is no excuse that the American construction industry has continued to build flimsier and flimsier houses out of traditional materials when technology itself is capable of providing new materials with finer insulation and soundproofing qualities. As stated by the Committee on Environmental Quality:

"Many of the old-fashioned dwellings of 40 or 50 years ago, by virtue of their more massive construction, larger rooms, numerous doors, hand operated appliances, and heavy sound-absorbent furnishings, were comparatively quiet places in which to live. In contrast, the modern dwelling with its light-weight construction, open plan design and multitude of noise makers provides very little protection from noise generated within or intruding from the clamorous outdoors.

"Although the building industry takes pride in its achievements, the fact remains that conventional building techniques have produced some of the noisiest buildings in existence. As a consequence, there is a strong and increasing public demand for noise control action in residential housing, particularly in multi-family dwellings."

Major property management firms report that noise transmission is one of the most serious problems facing managers and owners of apartment buildings throughout the country. They admit that market resistance not only is increasing as a result of excessive noise transmission, but that lack of both acoustical privacy and noise control is the greatest objection to apartment living. Similar concern is expressed by owners of hotels, motels, and office buildings which lack adequate sound insulation and noise control. Now, even the individual homeowner is protesting about the excessive noisiness in his home.

Dr. Robert Newman of the acoustical engineering firm of Bolt, Beranek & Newman, spoke to the same point in 1969. "In modern buildings, as we all know," he said, "there are many problems that were automatically solved in 'old fashioned' buildings, where we had heavy concrete construction, masonry partitions. We had a lot of mass, and as you know, this stops the transmission of airborne sound and reduces impact sounds as well. Today with lighter weight construction, we have long spans in buildings and we have high velocity air distribution. Then we have prefabricated components for partitions, for floors, for ceiling systems, all of these things have to be joined together at the site. They may be beautifully made, but unless they're connected together with completely air tight connections, we don't have any kind of sound isolation. Every leak and crack is a disaster, and the providing of air tight construction is terrifically important. We've got to recognize how these things sometimes interact with each other."

The 1968 study conducted by the Federal Council for Science and Technology listed seven fundamental causes, from a technical and economic point of view, for the rising noise problem in home and apartment buildings:

1. Increasing use of noisy high-pressure heating, cooling, and plumbing systems, power plants, and automated domestic

appliances. Progress in mechanization is outrunning advances in machinery noise control.

2. Poor acoustical design: open-space layout without regard to separating noisy areas from those requiring quiet or privacy.

3. Light-frame construction: increasing use of thin wall and floor construction and hollow-core doors which are poor noise barriers.

4. Poor workmanship: careless work by builders in sealing holes, cracks and noise leaks, and installing equipment.

5. High rise buildings: greater concentration of families in smaller areas result in noisier indoor and outdoor environments and increasing interfamily annoyance.

6. Higher costs of sound-insulated construction. The increased cost of constructing a sound-insulated building might range from 2 percent to 10 percent of the total cost of a building, depending upon geographic area, labor market, and other economic factors. Builders and owners of buildings are in a highly competitive market; therefore, they are reluctant to adopt new features which may result in higher building costs or jeopardize their competitive positions and profit margins.

7. The lack of mandatory acoustical criteria and enforcement. Until acoustical criteria are made mandatory and enforced by law, builders will continue to ignore them.

Although cheap houses with expensive price tags may be the norm today in America, all of the fault cannot be placed upon builders and developers. A major share of the problem lies within the antiquated building codes of thousands of cities and counties across the land. Because of past practice, for example, home builders almost without exception are required to build walls with two-by-four studs covered with thin plasterboard on each side even though use of new materials such as plastics in combination with paper honeycomb might provide a better and perhaps cheaper wall. Aside from the materials and construction methods, which seem to be generally locked in from ancient days, the United States is one of the few major nations in the world which does not have regulations setting minimum requirements for the insulation of buildings against sound. The United States, in fact, has no building code although rules established by the Federal Housing Administration are often considered to have weight similar to law.

The inclusion of noise-abatement requirements in building codes began in Europe as early as 1938. Since then most progressive countries have incorporated such criteria in their building laws, particularly sound-insulation requirements for multifamily dwellings. Most contain requirements or recommendations for both airborne and impact sound insulation. Some of the nations now enforcing rigid sound rules in dwelling construction are Austria, Belgium, Bulgaria, Canada, Czechoslovakia, Denmark, England, Finland, France, Germany, the Netherlands, Norway, Scotland, Sweden, Switzerland, and the U.S.S.R. The codes of Germany, England, Czechoslovakia, Russia, and Scandinavia are similar in the control of airborne sound, limiting transmission to 20-35 dB in the 100-Hz octave and ranging up to 50 dB at 3200 Hz. For impact noise, such as dropping shoes on floors or pounding water pipes, four of the countries limit transmission to 70 dB at 100 Hz, tapering down to 40 to 60 dB at 3200 Hz. In America, the FHA has issued a *recommended* curve for limiting impact noise in floor-ceiling construction in multifamily residences. The curve ranges from a permissible 66 dB from 75 to 600 Hz, down to 47 dB in the 2400-4800-Hz sound spectra.

Judge Theodore R. Kupferman, of the New York Supreme Court, in 1969 urged that federal legislation be passed specifically to control noise in the construction field. "We have found," he said, "that local jurisdictions are not in a position to cope with the problem at all. They've got so many problems they can't think of this, they don't have the funds to get involved.

"For example, in New York City, it took four years to get through the city legislature the new building codes that were going to have any provision at all for sound levels. And in the meantime, the East Side of New York, in the high rent area they are building the noise slums of the future, where, because they could have the proper tensile strength with the new materials, they could put up thin walls and it would stand 30 or 40 stories without worrying about the walls collapsing, but nothing was being done about keeping the noise out.

"In the old days we had ads in the New York subways by the plasterer's union saying if we talked to the world should your neighbors listen, in an effort to get us to use more plaster. But nothing was being done at the time, so finally they got a new building code put through. And how, if New York City is so far behind, can you expect it to be done in localities around the country?"

Suffice it to say that most homes and apartments today are not built with adequate safeguards against either indoor or outdoor noise, that which is airborne or carried through the structure itself. However, techniques and materials are available with which a quiet home may be built, or at least quiet rooms may be isolated from noisy ones. In general again, sound moving from one point to another may be softened by (1) acquiring a quieter source; (2) preventing transmission of vibration to adjacent surfaces; (3) placing a barrier, such as a wall, between the sound source and the listener, and (4) absorbing airborne sound with porous materials.

In a home or apartment, a basic consideration is the sound transmitted through walls, windows, and floors; in essence, the heavier the material used, the greater will be the transmission loss through a wall, window, or floor. As a rule of thumb, sound transmitted through a rigid partition decreases about 4 or 5 dB each time the weight of the material is doubled. Stated another way, if a wall is built of materials weighing 10 pounds per square foot, the noise passing through it would be almost twice as loud as if it were built of 20-pound-per-square-foot material.

This basic concept of mass or weight provides some interesting contrasts when translated into basic materials used in homes and apartments. For example, a wood door with ⅜-inch panels will provide 15 dB of sound attenuation, but a solid oak door 1¾-inch thick reduces sound 5 dB more. A ⅛-inch glass windowpane reduces sound transmission by about 27 dB, but twin panes of glass with an air space between provides 40 dB. A wall consisting of two ½-inch sheets of gypsum wallboard will give a transmission loss of about 30 dB, but a plaster wall on 2 x 4 studs provides 35. An excellent wall, again following the rule of increasing mass, may be built of 4-inch cinder block with plaster on each side. This provides a sound transmission loss of more than 45 dB. The transmission loss increases to 55 dB when a wall is built of two layers of 3-inch cinder block with plaster on each side. Note that the weight goes up in these illustrations from about one pound per square foot for the thin panel door to about 80 pounds per square foot for the two layers of cinder block with plaster. So, mass in wood, glass, concrete, brick, or stucco is one of the things to look for if you are shopping for a quieter house or apartment.

While engineers labor to design quieter kitchen equipment, as exemplified by today's virtually silent refrigerators, a fair amount of sound attenuation may be accomplished by preventing vibration from passing from one room to another through the house structure itself. For example, although today's refrigerators are fairly quiet, they still operate with motors which can transmit vibration through the floors of a house if the appliance stands on a hard surface such as linoleum. This can be improved by placing small pads of sponge rubber or polystyrene under the legs.

The dishwasher, generally a less perfected and thus noisier piece of machinery, often is installed badly with metal parts of the machine coming in direct contact with the wood on the kitchen counter, and the counter in turn attaches to the wall of the house. Thus, the vibration from the dishwasher is transmitted directly to the walls and a high level of noise results. This can be reduced by placing a thin layer of sponge rubber or other resilient, porous material at all points where the machine touches wood or floor. Considerable sound reduction can be achieved by applying a layer of glass wool or similar material completely around the three hidden sides of the dishwasher.

Water pipes, carrying the sound from an open faucet under the floor and throughout the house, are often a source of irritating noise. This noise may be eased by wrapping the pipes in cloth or insulation material. Especially, resilient material should be used to caulk any point where the pipes pass through a wall or part of the house structure.

Although many old-fashioned homes were plagued by knocking sounds in the hot-water heating system, this was not as serious as the fan noise carried through the metal pipe ducts running throughout the house from the furnace or central air-conditioning unit. These pipes, in addition to the furnace noise, often serve as direct "speaking tubes" from one room to another.

This condition may be eased by wrapping the pipes with insulation material, which serves the additional purpose of heat conservation. A relatively new product on the market, and an example of what the building industry can do if it tries, is furnace heat pipe fashioned entirely of compacted glass wool. With a rough surface on the inside of the pipe, the glass wool serves as a fine sound absorbent.

In preventing noise from traveling from one room to another, it is essential to eliminate any openings through which the sound can pass. For example, a badly fitted or warped door with an air space around the edges or under the bottom will permit almost as much sound to pass through as if the door were not there at all.

The use of sound-absorbing fibrous materials is probably the most commonly used technique for controlling sound in the home. The effect is most noticeable when draperies, carpeting, and soft-textured furniture are moved into a new home. Other common materials are fibrous acoustic ceiling and wall tile, and rough acoustic plaster, which is now a common ceiling material, at least in southern California although it may be innovative in other areas. The two basic properties to be remembered about sound absorption are: (1) the total amount of surface upon which sound may be dissipated as heat and (2) the fact that for an acoustical material to function, the air must be able to move freely among the surfaces. Glass wool, as an example, is an excellent sound absorber (with most effective frequencies varying with thickness of the material). However, if it is used inside a tightly sealed wall, it will be less effective than if small holes are provided in the noisy side of the wall so that air may move among the fibers. Acoustic materials, either batting or tile, are most useful in reducing the total sound occurring within a room, such as a recreation room where loud music and noisy recreation are common. If your teen-ager, for example, practices a musical instrument or plays his hi-fi at 100 dB into the late hours of the night, one partial solution would be to line his room with acoustic tile. A multitude of construction techniques exist, such as floating walls and floors, to ease the noise problem. Most are costly, some not so expensive; but if all else fails, the light sleeper can equip himself with a set of fitted earplugs.

Millions of American homes and other buildings today are assaulted constantly not only by noise, but by the intrusion of accumulated atmospheric fumes from industry, automobiles, and aircraft. If the environment continues to deteriorate, as it has through the past three decades, the average family may be forced to the ultimate recourse of sealing their home against the movement of air from the outside.

... Hotels, Offices, and Schools

The principles and techniques discussed in relation to home and apartment construction apply equally well to the control of internal noise in hotels, offices, and schools, although different combinations of sound problems may exist. Hotel builders and operators, in fact, probably practice the art of sound control to the highest degree in our society because their business essentially is providing people a place to sleep. Economic factors are obvious and predominant. Anyone who has spent a night in a motel with a window facing a busy highway or railroad will not return there again. On the other hand, the large number of excellent, and relatively quiet, motels to be found around the nation's major airports is a testimonial to the construction techniques which are possible when the profit dollar is involved. When similar techniques, with noise control written in, become a part of city, town, and county building codes, everyone's private place will become quieter.

The quiet office, which now is becoming the rule rather than the exception, is another example of what can be done when the effects of sound on economic efficiency are considered. It has long been recognized that the office should be quieter than the shop or factory, because the office is where the boss sits to make important decisions. Another more subtle factor in the creation of more comfortable offices is that more women are hired for office work than anywhere else in the industrial environment. If a secretary is annoyed by noise, she will say so, and the one person who is more powerful than the boss in any organization is the boss's secretary. Putting such status symbols aside, the largest single contributor to noise control in the nation's office buildings is central air conditioning. It works two ways. First, in order for air conditioning to work efficiently, the building must be sealed against the flow of outside air, which automatically serves the function of sealing out sound as well. The second contribution of air conditioning is that the murmur of fans and air flow usually provides the right level of "white" or masking sound necessary for privacy of conversation in most office situations. It has been found that absolute silence, occasionally broken by the clack of a typewriter, is not the best office atmosphere. Most people are found to operate best when a gentle noise, of perhaps 30 to 45 dB is present, either in a meaningless full-spectrum pattern or low music.

"We've come to realize that some noise is probably a very good thing," Dr. Newman commented recently. "Masking noise is a contributor to privacy, a contributor to quietness in many situations, especially in today's buildings where we're using lighter weight materials, using dry connections and where we have all the air leak problems. Many a time we find ourselves improving privacy situations by just upping the background noise level a little bit."

One key purpose of background noise in modern office buildings, where it is becoming the fashion to modify or do away with the private office where executives are wont to hide, is to mask private conversation from people who should not hear it. Conversely, the noise level must be low enough throughout to permit normal close-range conversation and use of the telephone. As a guide, originally formulated by Beranek and published by the Public Health Service, two people standing six inches from each other can barely hear each other in normal conversation if the noise background is 71 dB. If it goes up to 89 dB, the speakers must shout. A 51-dB noise level will barely permit audible normal conversation at a 5-foot distance, while a 69-dB level would require shouting. For satisfactory telephone use in the office, the maximum permissible sound level is about 60 dB, preferably lower.

The modern office typewriter is extremely quiet compared with typewriters in common use fifteen years ago and normally is no longer an annoyance or speech-interference factor in the office. And good secretaries may well reject a completely silent typewriter because they can no longer hear the speed and efficiency of their work. Some new business machines are noisy, including the instant copying machines and modern computer machinery, but in most cases it is possible to isolate this equipment within sound-absorbing enclosures so that the general office is not disturbed.

Office noise control is best achieved by proper architectural layout and arrangement, which, if done right the first time, can save tremendous amounts of money and materials in trying to fix after the fact. If offices are part of a factory building, they should be separated from the factory by spaces such as corridors, rest rooms, and storage rooms. In a large room where a number of people are working at desks, noise interference can be eliminated by appropriate desk groupings and the use of partial partitions between desks. Care also must be taken to prevent the direct transmission of private office conversation through heating or air-conditioning ducts or air-return channels.

Dr. Newman cites as an example a new science building recently built at the Massachusetts Institute of Technology. All offices and laboratories were provided with doors.

"In the drive for economy to reduce the cost of the building," Dr. Newman said, "it was decided that rather than return the air through a ducted system, air would be returned through the corridor, through louvered doors." The result was that occupants of all the offices and laboratories were constantly annoyed by sound drifting in and out from the corridors.

"So we screw up pieces of plywood and then put in transfer ducts and do all kinds of expensive things that could have been solved in the first place if we had just thought a little bit about the consequences. Every decision we make about a building has some kind of acoustical consequence. The selection of materials, the selection of finishes, the interrelation of elements, noisy and quiet, all of these things are going to affect the outcome."

Acoustic engineering in our schools is just as important as in the home and office. In schools, as in the office, the objective is not complete silence but a level of low noise conducive to study and work. In most classrooms, this is provided by the restless rustling of bodies, and teachers are generally tuned to the equally ominous onset of complete quiet or boisterousness. However, some basic factors must be considered in schools, particularly the prevention of sound spilling from one classroom into another.

Dr. Newman relates the story of an industrial high school built in an eastern city. It contained wood shops, metal shops, masonry shops, automobile shops, and offices, but for purposes of lighting and ventilation, the architects designed the building with concrete floor and a 13-foot-high concrete ceiling. Partitions between shops and offices extended only seven feet high.

"You have the cafeteria, the library, the principal's office, the lobby, everything all in one glorious space," Dr. Newman said. "It doesn't take any kind of imagination to conceive what this place sounds like. The architects said that they hoped that maybe the acoustics problems would go away. They didn't. The introduction of sound-absorbing material has helped in this situation, but some good old-fashioned partitions that have a little bit of weight in them, and that go right up to the ceiling, and some doors that can be closed, would solve the problem, but this can't be done. So they're stuck forever with a noisy great open space."

It is evident that our acoustic problems will not just go away, and the noisy great open space now encompasses our cities, our highways, and, in particular, communities near our jetports.

CHAPTER 2

In a silent African land
of no machines,
men and women retain good hearing
and health to advanced age—
and with no ulcers
or high blood pressure.

A
Quiet
Place . . .

THOUSANDS OF MILES and thousands of years of so-called civilization stand between the electronically amplified guitar in America and the primitive five-string lyre which provides background music for the festivals of the Mabaan tribe in Central Africa. Yet the civilization which spawned the electric guitar exists in the same moment of time with the Mabaans. Music, indeed, may be the linking symbol of contrast between a civilization which, on the one hand, enjoys excellent health and acute hearing to advanced old age while in America millions of young people are being deafened before their time by sounds of their own affluent pleasure.

Although the transistor radio has existed for several years now as a semipermanent appendage to teen-age ears, it is only recently that doctors have begun to express alarm and document the harm which is being caused to young ears exposed to long hours of rock music played as loud as 120 decibels in the discothèques of the country. While this relatively new danger to human ears is being explored, the story of the Mabaans is an outstanding example of modern research casting new light of understanding upon the roles of noise and aging upon the gradual loss of hearing acuity.

. . . The Mabaan People

The Mabaan people might well have gone unnoticed in their quiet corner of the world for more years if it had not been for Dr. Samuel Rosen, who, in addition to teaching at Columbia University's College of Physicians and Surgeons, is consulting otologist at Mount Sinai Hospital and the New York Eye and Ear Infirmary in New York City. Dr. Rosen, along with other hearing scholars, has long recognized that previous studies of hearing loss among aging people provided inaccurate results to an unknown degree because of the exposure of most people to the noise of modern civilization. Dr. Rosen therefore set out to study the hearing of a noise-free population to assess more accurately the effect of aging on hearing.

He learned of the Mabaans, about 20,000 people who live in a remote area of the Sudan, and organized an expedition to conduct a survey in the Central African area in December and January of 1960-61. Collaborating with Dr. Rosen in the work were Dr. Moe Bergman, Director of the Speech and Hearing Center, City University of New York; Dr. Dietrich Plester, associated with the Hals-Nasen-Ohren Klinik, Medizinische Akadamie, Düsseldorf, Germany; Dr. Aly El-Mofty, Professor of Otolaryngology, Ein Shams University, Cairo, Egypt, and Dr. Mohamed Hamad Satti, associated with the Stack Medical Research Laboratories, Ministry of Health, Khartoum, the Sudan.

The area chosen for the study is about 650 miles southeast of Khartoum, capital of the Republic of the Sudan, a few miles from the Ethiopian border and ten degrees above the equator. Until 1956 the area had been "closed," untouched by any foreign culture or civilization. The Mabaan country is primarily bush surrounded by swamps of the White Nile. It is accessible only during the dry season by truck or jeep over a narrow, rough dirt trail. The Mabaans are pre-Nilotic, pagan, primitive, tribal people whose state of cultural development is the late Stone Age.

"They are a peaceful and quiet people," Dr. Rosen described them, "living in small huts with straw-thatched roofs and bamboo sides, about eight to 10 feet in diameter. They have no guns, but hunt and fish with spears. They do not use drums in their dance and song, but pluck a five-string lyre and beat a log with a stick.

"The Mabaans are a shy, dignified people who shun contact with strangers. Their average adult appears to be about the same height as adults in our own culture. They are all lean, appear well-nourished and have a well-developed musculature with graceful, upright carriage. No obesity at any age was observed. The pigmentation of the skin is deep black, the lips do not protrude, the noses are broad and straight and the teeth are glistening white. They are alert, energetic and move about with quiet grace and self-confidence. For the most part they have been predominantly naked but are now beginning to wear some clothing."

Dr. Rosen reports the Mabaan diet is monotonous and almost free of animal protein since the people rarely slaughter their animals. They depend on trapping guinea fowl, rodents, and wild game, but the difficulties of capturing enough food with primitive equipment makes their animal protein intake very small. The main food is a sour, fermented, soft, pasty bread or gruel made of millet seed called *durrah*. The durrah contains carbohydrate, smaller amounts of protein, and very little fat. The millet seed also is used to make beer which is consumed in large quantities at harvest time. During the six-month dry season, the Mabaans eat fish cooked with okra and oil extracted from the kernels of wild dates. They seldom eat citrus fruits, but have groundnuts (peanuts), corn, and dry wild dates. The doctors reported that in spite of the monotonous and apparently unsatisfactory diet, the people appeared to be healthy and showed no signs of malnutrition or vitamin or protein deficiency.

The Mabaans are subject to certain native diseases but Dr. Rosen found *complete absence* of high blood pressure, coronary thrombosis, duodenal ulcer, ulcerative colitis, acute appendicitis, and bronchial asthma. Many of these diseases are common to our frenetic society in the Western world. Tests of 541 subjects of all ages from 10 to 90-plus revealed that blood pressure of the primitive people remains essentially unelevated from childhood to old age, an ideal circumstance. In the American population of apparently healthy individuals, however, the blood pressure increases progressively with advancing age, especially after age 40. Another finding was that the blood pressures of the Mabaan men (both young and old) tend to be consistently lower than those of women. In our own population, the blood pressure of men averages higher than women up to age 45, after which the women's blood pressure becomes greater than men.

The sound level in the Mabaan villages averages below 40 dB (the level of very soft conversation), perhaps comparable to the quiet in rural America a half-century ago, except occasionally at sunrise, when domestic animals such as roosters, lambs, cows, or doves make themselves heard. During six months of the year heavy rains occur about three times a week with one or two loud claps of thunder. Some of the Mabaan men engage in some productive activities such as beating palm fronds with a wooden club. However, even in this work activity, the absence of hard, reverberating surfaces, such as walls, ceilings, floors and hard furniture, apparently accounts for the low intensity of sound. This was measured at about 73–74 dB at the worker's ear.

The highest noise levels encountered by the five investigators during their stay in Central Africa occurred when the Mabaan villagers were dancing and singing. One recorded group consisted of ten young men and ten young women. Stanzas of a song were sung by a very soft-voiced male who also played the five-string lyre, followed at the end of each stanza by the chorus of twenty. The recorded levels of twenty singers in chorus were 100 to 104 dB, topped by hoots and shouts at the end yielding levels of 106 to 110 dB. Such festival singing apparently occurs about one to three times a week and lasts from one to three hours. For the most part, this activity occurs over a two-month period celebrating the spring harvest and very little occurs during the six months of the rainy season.

. . . Results of Hearing Tests

So the Mabaans live in a truly quiet place, their ears and nervous systems finely tuned to the sounds of nature and disturbed by none of the machine and electronic noises found in the industrial societies of the world. The five investigators, using carefully calibrated equipment, measured the hearing of the 541 subjects. They were gathered by interpreters and other assistants in a village and transported to the test area in a large lorry. No effort was made to select test subjects. All available men, women and children—ranging from 10 to 90 years of age—were brought in for the tests without attention to occupation or other special factors. The hearing of men, women, and children was measured in six sound pitches, or cycles per second, ranging from 500 to 6000 Hz.

Among 174 men and boys in the 10-19 age group, the hearing threshold ranged from 21 dB at 500 Hz to 26 dB at 6000 Hz. (This range is roughly comparable to the level of background noise permitted in a very quiet radio broadcasting studio.) As the tests ranged upward in age, the men showed very slight losses. In the 40-49 age group (106 subjects tested) the average hearing threshold at 500 Hz was 24 dB, rising to 35 dB at 6000 Hz. The twenty men who were tested in the 70-79 age range had median threshold of 26 dB at 500 Hz and 41 dB at 6000 Hz. Thus, even at the threshold of age 80, Mabaan men are able to hear quiet normal tones of conversation. Opposite to the patterns normally found in industrial societies, the hearing of Mabaan women in their later decades was less acute than that of the men, but still much better than their counterparts in America or Europe.

The contrast was seen more clearly when five investigators compared their African results with the findings of several thousand hearing tests which were conducted with a random population sampling at the Wisconsin State Fair in 1954. The comparison showed that in every decade of age change, the Mabaan hearing loss was much lower than that among the Wisconsin subjects. Comparing only the most severe hearing decline, the 70- to 79-year-old men at 6000 Hz, the Mabaans averaged a loss of only 15 dB, the Wisconsin subjects nearly 65 dB.

The Wisconsin Fair study shows a sharply increasing loss of hearing with each decade of life, particularly at the high frequencies but clearly involving all test frequencies. In contrast, all the Mabaan groups divided in ten-year spans from the 10-year-olds through the 79-year-olds fall within the first two decade lines of the Wisconsin group. Obviously, the hearing of the Mabaan men is retained dramatically with aging in comparison with the sharp losses suffered by the Wisconsin men. The average 80-year-old Mabaan man hears as well as an average 30-year-old resident of Wisconsin.

"It is tempting to speculate on the reasons for the striking differences," Dr. Rosen commented. "The Mabaan environment is dramatically quieter at almost all times than environments of populations sampled in our culture. The only high-level noise recorded in our stay in the village was during the merrymaking of the youths when they sang and played their crude musical instruments. Except for occasional transients, usually the fleeting noises of domestic animals, few other sounds were sufficiently intense to yield a reading on the sound level meter, which could read as low as 34 dB. Other researchers make a case for noise as the critical factor in the differences in hearing with aging in various populations. The results of the Mabaan study appear to support this view."

More than a year later in 1962, Dr. Rosen and Mrs. Helen V. Rosen, Dr. Plester, and Dr. El-Mofty returned to the Sudan to test Mabaan hearing in the higher frequencies ranging up to 24,000 Hz. Most hearing surveys have indicated that most Americans above age 25 are seldom able to hear sound about 16,000 Hz and the new tests bore this out. However, in order to broaden the range of comparison, the investigators this time compared the Mabaan findings with hearing surveys in New

York, Düsseldorf, and Cairo. The tests were conducted at 12,000, 14,000, 16,000, 18,000, 20,000, 22,000, and 24,000 Hz.

In the 10-19 age group, 94 percent of the Mabaans heard 18,000 Hz, compared to 88 percent of the combined city populations. Fifteen percent of the Mabaans heard 22,000 Hz, but only 6 percent of the city populations did. At 24,000 Hz, 10 percent of the Mabaans still could hear, but only 1 percent of the young city dwellers could detect the test tone.

As the tests progressed upward through the age groups, even greater differences emerged between the Mabaans and the residents of the three large cities. In the 40-49 age group, 57 percent of the Mabaans could hear the test tone at 16,000 Hz which was audible only to 26 percent of the metropolitan subjects. At 20,000 Hz, 4 percent of the Mabaans could hear, but none of the city dwellers.

Between ages 50 and 59, the Mabaans had a median hearing threshold of 67.5 dB at 14,000 Hz, but no median was obtained for any of the urban groups since there were too many who did not respond even at 90 dB. In the 60-69 age group, 99 percent of the Mabaans could still hear 12,000-Hz signals, compared to 32 percent of the city populations. Up to age 79, 53 percent of the Mabaan people still could hear 14,000 Hz at 91.8 dB, compared to only 2 percent of the city subjects. At 16,000 Hz, 5 percent of the Mabaans heard, but none of the others.

The Mabaans in their calm, quiet, bucolic existence unquestionably have demonstrated better hearing to advanced age than their counterparts crowded into noisy cities of industrial society. In examining the question of why long exposure to increasing levels of sound causes ears in our society to age faster, the effect of acoustic load has been compared with the wear on a rug covering a flight of stairs in a house. The rug represents the set of hair cells within the cochlea; people climbing the stairs represent the movement of the inner ear fluids. High tones, as it were, use the rug of the lower floors only, whereas the low tones march all the way up to the top of the stairs; thus, wear on the rug is greater on the bottom cells than upon those near the top.

"Based on this concept," said Dr. Rosen, "presbycusis, the loss of hearing with age, may be regarded as a kind of degenerative disease due to wear which starts early and intensively in civilized populations with their noise-laden environment . . . sounds impinging upon the ear represent the greatest mechanical and metabolic strain for the hair cells of the basilar membrane of the basal coil of the cochlea."

Another clue that part of the hearing loss normally ascribed to aging is actually due to noise comes from the observation that the hearing of males ages faster and earlier in our culture than the hearing of females. In most instances, men are professionally exposed to higher levels of noise than women. Among the Mabaans, however, where men and women are equally exposed to the same environmental noise, the hearing of females is not significantly better than that of the males.

"Why is the hearing in the higher frequencies of these primitive people excellent through life (ages 10 to 80)," asks Dr. Rosen, "whereas in modern industrialized areas in the United States the hearing in the higher frequencies tends to deteriorate in the natural course of aging? In addition to the effects of noise in our culture, high blood pressure and atherosclerosis of the small blood vessels to the internal ear may contribute to this effect.

"It seems that in Western civilization there are factors other than noise which contribute to presbycusis and degenerative changes with aging. The Mabaans age more slowly than we do. They have no vascular hypertension, coronary artery disease or duodenal ulcer. They have little atherosclerosis, no allergies or bronchial asthma, and have extremely little stress and strain in their lives. . . . Might not the stress and strain which afflicts modern civilized man somehow affect all his senses, including hearing?"

It is already possible to predict what will happen to the Mabaans as they progress from the simple primitive to a more modern and complex society. According to the Chief Internist of the Khartoum Civil Hospital, Dr. Abdul Mohamed Halim, the Mabaan people become prone to high blood pressure and coronary thrombosis when they move to north Sudan, where they are exposed to noise, different diet, and the stress and strain of city life.

Noise pollution won't kill you.
It can only drive you nuts or make you deaf.

There's so much noise in this city that you'd hardly think there's a need to advertise it.

Noise, after all, is something New Yorkers work with, eat with, try to sleep with, and wake up with—usually when they don't want to.

New York's noise is so pervasive that, except for those few noises loud enough to make themselves heard above the general din, it's usually taken for granted. And it shouldn't be.

Noise is dangerous.

If you think about it, you'll realize that most of the dangerous things in this world are noisy.

A gunshot is dangerous. It's also noisy.

Breaking glass is noisy. It's also dangerous.

As a result, you instinctively react to noise as a warning of danger. And living with the noise in New York is like having someone fire off a pistol behind your back 24 hours a day.

But while consciously you may be blasé about noise, your body, your nervous system and your subconscious are naive enough to keep right on reacting to it.

It can drive you nuts.

Loud, unexpected noises are very good at creating emotional stress. When the noise is continual—as it is in New York—so is the stress.

And continual emotional stress, as many psychiatrists believe, is enough to turn a normal adult, with normal problems, into a neurotic—if not an out-and-out psychotic.

In Ohio, for example, a scientist was so maddened by Air Force bombers flying over his house in the middle of, the night that he tried to shoot them down with a rifle.

And in Japan, the noise from a piledriver made a college student find some peace and quiet by sticking his head between the pile and the descending hammer.

If you've been living in the city for a number of years and you haven't gone off the deep end yet, you're still not off the hook. Because if New York's noise pollution hasn't affected your mind, it's probably affecting your hearing.

It can make you deaf.

Ever since the last century, when blacksmiths and boilermakers started complaining to doctors about their hearing, the medical profession has known that noise can produce deafness.

Today, subway conductors, jackhammer operators and factory workers, to mention just a few, earn their living at the risk of their hearing. Unfortunately, millions who don't work in these occupations are subjected to the same occupational hazards.

Portable air compressors on New York streets are loud enough to drown out dynamite blasts.

Food blenders in our kitchens are actually louder than Niagara Falls.

Power mowers and poorly-muffled motorcycles are so noisy, they make factories seem like libraries.

As the noise gets louder and louder, the people who have to live with it get deafer and deafer. It used to be that people didn't start to lose their hearing until the age of 70. In big cities today, people start going deaf at 30.

If we don't solve the noise pollution problem now, in a few years it could automatically solve itself. By making us all too deaf to hear it.

Nobody ever does anything about anything unless people demand it.

In a big city like ours, individuals with legitimate complaints are dismissed as nuts, lunatics and crackpots.

But if enough "nuts", "lunatics" and "crackpots" get together, there's no telling what they can accomplish.

In Westchester, enough people were sick and tired of having their sleep interrupted by trucks with roaring exhausts to get a noise limit set on the New York State Thruway.

In Florida, enough people were bugged by noisy electrical appliances to get a law passed against them.

And in New York, enough people were infuriated by the noise they had to live with in their apartments to get at least some soundproofing written into a new Building Code.

But while there are enough concerned people to get some of New York's noise shut out of future apartments, there are still far too few to get it shut up for good. That's why there's such a thing as the Citizens for a Quieter City.

Your financial support is welcome. Your moral support is essential.

Citizens for a Quieter City needs your money to make people aware that the problem of noise pollution exists. And that it doesn't exist as something to be passively accepted as an inevitable burden of the human condition.

Your money will also help stimulate research on the problem and its solution.

But what's even more important to us than your money is your name.

With the names of twenty or fifty or a hundred thousand New Yorkers who are as outraged by noise as we are, we can prove that this city has had all the noise pollution it's going to take.

So please fill out the coupon. And don't feel embarrassed about sending it in without any money. A lot of people's names could be more help to us than a few people's money. Because the one thing we can't afford to do about noise pollution is keep quiet about it.

Public service ad designed for Citizens for a Quieter City, Inc., by Scali, McCabe, Sloves, Inc. Reprinted by permission.

CHAPTER 3

*Rock, electronically amplified
to unbearable levels,
deafens a generation of young people
before their time.*

The Sound of Music . . .

WHEN THE MABAANS have demonstrated that a quiet land and a quiet way of life are so beneficial to the organs of hearing and health in general, it seems a pity that we have allowed the nerve-shattering sounds of our machines to impair the quality of our environment. The shame is compounded when we deliberately expose ourselves to ear-damaging sound in the name of pleasure, and this is happening to an alarming degree among millions of young people.

Although rock music (or whatever name the newest musical

sound may be wearing these days) is only good or bad according to the quality of the musicians, it is bad per se when the intensity is so great as to cause hearing loss. For a number of years this potential result has been the subject of wry jokes among long-suffering adults who thought the sound with its extreme amplification was *their* problem. Now we are learning the youngsters themselves are suffering from the electric guitar and all of the other electronic monstrosities it has spawned in music along the way.

Rock and roll music has been measured at peaks of 122 dB in discothèques and teen dance halls, where reflecting surfaces cause reverberation and intensify the sound until a steady state is reached far above the 80 to 90 dB of noise which has long been established as the risk criterion level in industry. To make the matter more critical, much of the highest intensity music occurs in the range of the human voice and thus, first of all, threatens loss of hearing acuity in the range most needed for understanding speech.

. . . Electronic Amplifiers

The fault with the music is not so much the instruments themselves, or the sustained high level, but with the electronic amplifiers. An old-fashioned brass band playing a Sousa march on Saturday night in the town square generated as much sound, but it was not amplified and it dissipated in the open air. A few of the musicians might have had a tin ear for a few hours after the concert, but the temporary threshold shift soon passed away. But the modern sound, in addition to the electric guitar, includes electronic amplification of virtually every known musical instrument including drums. The game among the thousands of rock groups here and abroad is to see which group can afford the most massive amplifiers and thus, apparently, drown out the sound of the competition.

"With this situation," comments Dr. Robert Feder, Beverly Hills ear specialist, "everything is reamplified many times and the noise becomes nearly intolerable." Intolerable is the word adults use, but the teens who enjoy it indicate pleasurable obsession with the *feel* as well as the sound of the intense vibrations.

Dr. Charles Goodhill of Hollywood reports sound levels in many rock clubs up to 125 dB. Dr. Charles P. Lebo of the University of California measured sound levels in two San Francisco rock joints and found sound intensity averaging 100 dB in virtually all frequencies with a peak of 119 in the center of one of the halls. Dr. Lebo estimates that 10 percent of the people in such a hall would show no hearing loss effects, 80 percent might have their hearing threshold raised by 5 to 30 dB, and 10 percent would suffer temporary threshold shift of 40 dB. As for permanent damage, permanent threshold shift varies broadly according to intensity, the length of time of exposure to sound, and whether or not it is continuous. Some researchers feel a steady diet of rock music for a week would cause permanent hearing loss. No one has been able to show the loss statistically in the case of youngsters who listen perhaps to three hours of music a night—in a dance hall, on radio, or on record—but are not exposed otherwise.

"Essentially," Dr. Lebo says, "the aging process accelerates so that 20-year-olds have 60-year-old ears. Usually sound is dissipated in open air, or in a living room or at the symphony, absorbed by furniture and rugs. But in a discothèque or one of these big empty halls, there's no place for the sound to go and it is reamplified over and over."

Earphones

He warns that some hearing deterioration can be produced by regular listening to high-intensity sound by use of earphones. "The earphones are great to use, of course," according to Dr. Lebo, "and a blessing to everybody else in the house but they are dangerous because the sound is delivered directly to the ear without any kind of muffling." Although the statistical proof may yet be lacking, Dr. Lebo worries that considerable permanent hearing damage may be resulting from the rock music mania. "The hearing loss, which largely affects the ability to understand speech is caused by injuries to the nerve endings and is called 'nerve deafness,' " he pointed out. "Though doctors can repair ear bones and rebuild eardrums, this kind of injury is beyond us.

"Rock musicians I consulted feel that some performers are merely exploiting the parameters of amplification now and will eventually revert to more bearable limits." Until that happens, however, Dr. Lebo recommends earmuffs or plugs. "When my children go to concerts," he said, "they take along little ear protectors. I only hope they wear them."

Dr. George T. Singleton, an ear, nose, and throat specialist at the University of Florida, noticed that when he picked up his teen-age daughter after a dance she couldn't hear what he said in the car on the way home. Singleton set up a research project and tested the hearing of ten 14-year-old youngsters an hour before their next dance. The investigators then tested the dance hall, where they found the sound pressure level at 106 to 108 dB on the dance floor. The test crew had to move 40 feet outside the building before the sound level dropped to 90 dB. After the dance, the ten test subjects showed an average temporary threshold shift of 11 dB and one boy's ears had been dulled by 35 dB. As for the musicians themselves, most are not in the direct projected path of the amplified music, but as the reverberation reaches a steady-state level in a closed hall, the instrumentalists suffer. Dr. James Jerger of the Houston Speech and Hearing Center tested the hearing of a five-man combo. One player had a 50-dB temporary loss of hearing, and three had already suffered a slight but *permanent* loss. All of the men were younger than 23.

Although these scattered returns may give some indication of what is happening to what may become known as the deaf generation, a number of investigators have set out to gather clinical proof and experimental evidence of what actually occurs inside the inner ear. Probably foremost among these is Dr. David M. Lipscomb, assistant professor and Director of Audiology Clinical Services in the Department of Audiology and Speech Pathology at the University of Tennessee in Knoxville.

High-Frequency Hearing Impairment

Once considered a rare phenomenon," Dr. Lipscomb commented, "the incidence of noise-induced hearing impairment in children from exposure to high-intensity sounds in their play and recreation environment is now not only common but increasing at an alarming rate." He noted results of a hearing test conducted with 1000 Colorado children which showed high-frequency hearing impairment to be more frequent in older than in younger children and more general among boys than girls. In 1968, Dr. Lipscomb and his associates tested the hearing of 3000 youngsters in the Knoxville public schools, 1000 each from the sixth, ninth, and twelfth grades. The screening level used was 15 dB, lower than the standard 25-dB signal used for hearing tests, for the purpose of detecting even minor hearing losses. Among the sixth graders, high-frequency impairment was detected only in 3.8 percent, but this jumped dramatically to 11 percent among the ninth-graders. High school seniors showed a 10.6 percent incidence of HFI. "We had expected some increase in the incidence of high-frequency hearing impairment as a function of age," Dr. Lipscomb said, "but we did not anticipate such a marked rise. We concluded from this survey that young people may be losing some hearing because of the high-intensity sound environment to which they subject themselves. A common source of such high-intensity sound is rock n' roll music. Discotheques and teen dance halls contain dangerously high-intensity sound."

Dr. Lipscomb then measured the sound in several dance halls. His findings corresponded closely with those of Dr. Lebo, previously mentioned, with sound levels ranging from 104 dB at low frequencies, up to 120 dB in the 125- to 250-Hz octave bands, and never lower than 100 dB in the frequencies from 500 to 8000 Hz. These findings in turn were compared with the damage-risk criteria established as noise control safety orders in industry in 1962 by the California State Department of Industrial Relations. These orders require that noise not exceed 110 dB in the low frequencies and never rise above 95 dB in the 500- to 8000-Hz octave bands. Obviously, the sound level in the teen dance halls was consistently higher than that permitted in industrial factories in California.

Carrying his research a logical step farther, Dr. Lipscomb set up an experiment, using guinea pigs, to measure the extent of anatomic damage of the hearing mechanism by intense sound stimulation. The guinea pigs were subjected to rock music approximating the levels measured in the dance halls. Before subjecting the animals to the noise environment, each was lightly anaesthetized and a plug was inserted into the left ear, as an experimental control, to protect it from the sound. The sound stimulation schedules were set up to approximate the random exposure by teen-agers listening to high-intensity music. Over a period of 58 days, the guinea pigs were exposed to the high-level noise 27 times, the exposures varying from a minimum of 35 minutes to a maximum of 227 minutes. After 65 hours of total exposure, each animal's protective plug was removed and then both ears were exposed to an additional 23 hours of rock music. Surgical specimens then were taken from the cochlear tissue of the inner ear and examined microscopically.

The tissue, taken one and one-half turns from the base of the organ of Corti in the protected ear, showed that all structures and cells appeared normal. However, tissue taken from the identical area of the unprotected ear showed widespread damage. Several inner hair cells were obliterated or misshapen and numerous outer hair cells were collapsed and appeared to be missing. Dr. Lipscomb found that up to 25 percent damage to hair cells was found in other sections of the cochlear coil.

"Most alarming," he wrote in Clinical Pediatrics of February 1969, "is the widespread irreversible damage in the cochlea of an experimental animal exposed to sound comparable in intensity to that which many young persons expose themselves every day. Caution must be exercised, of course, when relating observations from experimental animals to humans. The inference is clear, however, that the typical discothèque sound environment is sufficiently intense to be extremely hazardous to the health and well-being of sensory cells in the cochlea." Dr. Lipscomb also cautioned that the hair cell damage was so erratic and random in pattern that pure tone audiometry "cannot delineate early signs of damage resulting from sound trauma, unless the cochlear injury is specifically located in one of the sensory regions tested by the pure tone stimulus."

.. The Need for
Hearing Conservation

Persons who expose themselves to high-intensity rock music for a total exposure time exceeding twenty-three hours in a two-month period may suffer irreversible damage to the sensing cells in the ear, Dr. Lipscomb advised. He urged that a program of hearing conservation be initiated. The dangers of excess exposure to intensely loud music sounds must be made clear to our youngsters, probably by pediatricians who normally see young people and give advice on physical problems. Dr. Lipscomb recommended the use of earplugs. "With the use of imagination," he said, "these could be made quite attractive and stylish, perhaps to the extent that their use might become a fad among young persons.

There are those who could tell Dr. Lipscomb that if an adult suggests anything in this day of mental confusion and rebellion, it will not become a fad among the young. However, certainly rock music is here to stay although tomorrow's form may not be the same as today's. To give some impression of magnitude of the loud-music business, Columbia Records five years ago did 15 percent of its business in rock music. Today the company's rock productions constitute 60 percent of its entire output, which in total has been growing at 15 to 20 percent per year. Personal appearances of the top groups are booked all over the nation in small clubs, theaters with up to 2000 seats, and stadiums and arenas with 20,000 to 60,000 seats. The rock phenomenon is getting bigger, fed primarily by the more affluent young, and in turn producing a new breed of youthful millionaires. Some social commentators see rock as a major element in a series of manifestations among the young that may portend a cultural revolution—away from rationalism and puritanism, toward the senses—and the experimental modern music often is blamed for enticing the young to drug use. As with all intemperate use of the senses, however, this glut of extreme noise leads directly toward the dulling of a very vital sense.

What can be done about the extreme levels of amplified music? One school principal in the Los Angeles area has attacked the problem directly. Dr. Rodney E. Phillips, principal of the Earl Warren Junior High School in Solana Beach, said: "I used to go home from school dances with my ears ringing and severe headaches. When my wife said something to me, I couldn't understand her."

So now when he chaperones a dance, he carries a decibel meter with him and the level of the music is never (with one exception) allowed to exceed 100 dB. "I don't have any trouble enforcing it," Dr. Phillips commented. "It's written into the contract with the musicians, and I carry the contract in one pocket, the money in the other." Dr. Phillips does permit the band to play the last number of each dance as loud as it wants to. The kids cheer and the principal says he can feel the building vibrate by touching a wall.

Most experts are mystified by the fact that youngsters like their music so loud, because their ears, above all, are most sensitive to begin with. The pain threshold is 120 dB for some ears. At 140, sound can be extremely painful, and 160-dB sound can kill small animals.

It is a sad commentary that in our civilization, which already is highly restrictive, that more laws may be necessary to prevent our youngsters from losing their hearing acuity. However, two U.S. Senate subcommittees already are considering the advisability of conducting hearings on the rock n' roll question.

The saddest commentary, about this question as well as about all of our environmental pollution, is that the situation must become dangerous and even critical before anyone does anything about it.

CHAPTER 4

*Freeways shatter man's peace
with the roar of 100 million autos,
while mass rapid transit systems
that would shrink traffic
to manageable proportions languish.*

Freeway
Frenzy . . .

IN JULY 1968, the New York State Court of Appeals held that the state must pay a property owner for damages caused by the noise of passing traffic when the state takes part of his land to build a highway. The court awarded $37,000 to Ira and Dorothy Dennison after land in their remote wooded area in the Lake George region was taken for a highway interchange. In the court's majority opinion, Judge Kenneth B. Keating wrote: "Of particular concern has been the damage done by massive public highway construction to the quiet beauty of many once remote and inaccessible areas, as well as the intrusion of the seemingly endless lines of asphalt and concrete into the enclaves which many people have sought as surcease from the hustle and bustle of modern-day life."

The landmark decision was one of the first small chinks in the formidable armor which has developed around the public and private interests which each year devote approximately one-seventh of the entire gross national product to automobiles, trucks, and buses, and the massive ribbons of concrete upon which they must move. Of all the noise and other environmental pollution that assails the average American citizen today, that from automotive traffic is most general and irritating, indiscriminately invading our cities, our suburbs, and our once quiet countrysides.

The legal chink in the armor was a very small one, and conflicting decisions are likely to be rendered by many courts before there is any real hope for general relief from automotive noise. The New York Circuit Court, in fact, handed down a dissenting opinion written by Judge Francis Bergan. "Millions of residents of New York are exposed to noise along the Long Island Expressway and the East River Drive, to suggest some notable examples," he said. "That it has a consequence on market value wherever it is heard is undoubtedly true. But there are some unpleasant consequences of modern life which are not the proper subject of damages in a law court."

Almost a year later, in June 1969, this same opinion was expressed by the Utah Supreme Court, which held in a similar case that a home owner is *not* entitled to compensation for noise resulting from relocating a highway. In this case, the road was moved 37 feet closer to a man's home. He had been paid for a strip of land taken from him, but he sought additional money because of the noise. The court denied payment based solely on the increased noise.

It is paradoxical, but also typical of our industrialized and technical society, that the internal combustion engine which moves our freight and gives us almost unlimited individual mobility is at the same time probably the most serious polluter of our environment, both in air pollution and noise. As shown by the two court cases previously cited, this automotive monster has been permitted to proliferate with virtually no control of what it does to all of us. The paradox is deepened by the fact that while each of us abhors the fumes and noise of street and highway traffic, no one is willing to give up his auto or truck to make a better world for the rest of us.

In the words of Prof. J. Alan Proudlove, professor of transportation studies at the University of Liverpool in England, "The urban motorway is one of those unfortunate innovations which, while achieving the narrow objectives of its own technology, creates more difficult problems in associated areas. It has introduced a major disruptive element into the urban environment. These problems are those of scale; of visual or physical integration with the buildings adjoining the motorways, and of noise, which is probably the most intractable of side effects of the urban motorway, and indeed of increasing car ownership.

"We continue to acquire more cars and demand more and better roads and, as motorists, we go to great lengths to drive as close to our destination as possible. It is only when other people's cars threaten our own environment that we complain. We also need a much more liberal attitude toward compensation for loss of amenity brought about by traffic intrusion."

Professor Proudlove thus touches upon the two key points involving traffic noise: (1) vehicles are multiplying at an increasing rate, and (2) the cry for relief is only beginning to be heard. Oddly enough, in the beginning the cry for relief was louder than the vehicles themselves.

The complaints about road traffic noise, indeed, began at least as early as the eighteenth century when Schopenhauer wrote: "The truly infernal cracking of whips in the narrow resounding streets of the town must be denounced as the most unwarrantable and disgraceful of all noises. It deprives life of all peace and sensibility. Nothing gives me, so clear, the ground of the stupidity and thoughtlessness of mankind as the tolerance of the cracking whips. This sudden sharp whack, which paralyzes the brain, destroys all meditation, murders thought, must cause pain to anyone who has anything like an idea in his head. Hence, every crack must disturb a hundred people applying their minds to some activity, however trivial it may be, while it disjoints and renders painful the meditations of the thinkers, just like the executioner's axe when it severs the head from the body."

Shortly after the first steam-powered horseless carriage appeared in England, the British Parliament in 1831 passed the famous Red Flag law, which required a man to precede a horseless carriage on foot, carrying a red flag by day and a lantern by night. This law, in fact, was credited with slowing the development of automobiles until 1896, when it was repealed. Coincidentally, that year saw some of the early application of the internal combustion engine to vehicles in England, France, Germany, and the United States. When the explosive noises of the automobile disturbed the New England countryside, Vermont in 1894 passed an adaptation of the English law which was to be repealed two years later. To protect the nervous system of horses, and perhaps people's ears as well, the Vermont law required a man to walk "several hundred feet" in advance of a moving car.

By the turn of the century, however, the automobile obviously was beginning to win, and the victory was assured by Henry Ford's mass production of the Model T. The only concession to human ears was the invention of the muffler, which has been improved somewhat through the years from original designs, but the auto industry for the past thirty years has been involved with a fetish for removing each slight click and rattle for the benefit of the *occupant* of the automobile. Thanks to various improvements, the average new automobile today is relatively polite and quiet. However, we have achieved in turn a rather unhealthy worship of the private vehicle which manifests itself in powerful sports cars owned by people who obtain ego satisfaction from the powerful sound of their vehicles. The same syndrome infects the growing millions of noisy motorcycle owners. Despite laws against motor vehicle noise in most U.S. communities, the enforcement problem is virtually impossible, due to the immense number of autos, trucks, and buses on the highways today. Diesel trucks, especially, go roaring through the day and night disturbing the peace and rest of millions living within a mile or more of the freeway.

In addition to engine noise, the multiplication of vehicles on four- and six-lane roads causes levels of noise, of tires against pavement, which approach ear-damaging levels for residents living nearby. Nearly 100 million motor vehicles are in operation in the United States today. Eight to 10 million new ones join the throng each year while lesser numbers are consigned to the junkpile. Some estimates indicate the number of private vehicles, using more thousands of miles of paved highways, may reach 240 million in America by the turn of the century.

"Noise radiation from vehicular traffic is becoming a major source of complaint among urban and suburban dwellers," reported the national Committee on Environmental Quality. "Vehicular traffic, especially highway noise, is the most serious offender. Generally speaking, traffic noise radiating from the freeways and expressways and from mid-town shopping and apartment districts of our large cities probably disturbs more people than any other source of outdoor noise. Although air-craft noise is much more intense, the exposure time is sub-stantially less than that of round-the-clock, continuous highway noise.

"The generation and intensity of traffic noise is dependent chiefly upon (1) the kind, number and speed of vehicles; (2) the character of the vehicle roadbed interference, and (3) the type of environment—mid-city, suburban, rural—in which the problem exists."

Of all the types of vehicles traveling on our expressways the trailer truck is the most notorious noise producer. At expressway speeds, a single truck may generate sound levels exceeding 90 dB; while a long line of truck traffic may produce noise above 100 dB. Because expressway truck traffic generally is heavy during the night, when natural background noise levels are low, the noise seems much louder than during the daytime. This is one of the major reasons why traffic disturbs millions in their sleep and relaxation. Following trucks in a descending order of annoyance are buses, motorcycles, sport cars, and passenger automobiles.

A vehicle is a complex noise generator containing a multitude of sources. For example, the overall noise radiated by a truck might be generated by the exhaust system, the engine, the transmission, brakes, horn, tires and loose chains, pins and cargo. Although some effort has been made recently to reduce

FREEWAY FRENZY

the noise levels within passenger automobiles, very little has been done by truck and automobile manufacturers to suppress noise radiation by the vehicle itself, aside from some improvements in muffler design and, more recently, tire tread design. The current trend toward producing larger and more powerful trucks, raising speed limits on our expressways, and expanding the volume of truck traffic will greatly increase traffic noise unless effective countermeasures are taken.

To date, highway engineers and city planners have given little consideration to the effects of traffic noise on a community. They tend to excuse the routing of highways through quiet residential areas next to schools, churches, and hospitals or across peaceful recreational parks as "the price of progress." The design and location of the highway is usually dictated by such socio-economic factors as minimum route mileage, lowest land acquisition cost, safety, public benefit, and toll charges or other cost-defrayment or cost-allocation schemes.

While these are valid considerations, other important economic factors should be considered: namely, the value depreciation of property along noisy expressways and the high cost of sound-proofing apartments, schools, churches, and houses located near expressways. Efforts undertaken at the national level to control traffic and rail transportation noise are insignificant in terms of the scope, magnitude, and severity of the existing problem. Progress, in transportation at least, is *not* improving the human nervous system.

. . . Traffic Noise Levels

Millions of Americans today live with 100-dB noise levels where homes and apartments are built within 200 feet of a busy downtown street or expressway. The expressway or freeway indeed has carried these high noise levels and atmospheric pollution into suburbia and into the open countryside with the interstate highway system. Most people remain indoors most of the time and thus are protected from ear-damaging sound from road traffic, which, with a wall to attenuate it, may fall to 70 or 80 dB. In metropolitan areas, the level of traffic noise seldom falls below 70 dB, except during the hours from 2 to 6 A.M. The problem is not limited to America, by any means, but becomes acute in any country as its level of freight movement and individual mobility increases.

In Russia, for example, the government publication *Izvestia* recently declared that noise is hurting Moscow's health and affecting the productivity of workers. Biggest offenders are diesel trucks "grinding through the middle of the city at night." The editorial writer complained that "there still are citizens who, not taking into consideration the people around them, start car engines at night or drive around the courtyard on motorcycles with rattling and roaring noises." *Izvestia* called upon police "to punish such malicious violators of peace and quiet more strictly" and urged the passage of more stringent antinoise measures.

In London, a survey of residents showed traffic to be the most serious annoyance of all sound sources. The survey covered 400 different locations within the city, and traffic noise predominated in 84 percent of all complaints received. The noise climate ranged from a daytime high of 80 dB near arterial roads with many heavy vehicles and buses down to a low of 40 dB at night in homes situated a considerable distance from any traffic noise. Some traffic noise was measured at 90 dB and above.

According to Prof. William Burns of the University of London, the survey showed that the highest noise levels are associated with buses and heavy goods vehicles, with occasional motorcycles and sports cars. The heavy diesel-engined commercial vehicle is the main source of noise, and noise seems to be a built-in feature of these vehicles. In view of the similarity of design on a world-wide basis, this is an important technical problem with very wide consequences.

Although the diesel engine and unmuffled gasoline engine are the major offenders, tests also have been conducted to show the amount of noise produced by a modern automobile equipped with muffler. In one survey, two cars—one a low-cost American car, the other an upper-medium-priced auto—were checked on gravel road at 30 miles per hour, and upon a smooth asphaltic concrete road at 70 mph. The resulting noise ranged from 85 to 90 dB in the low sound frequencies (35 to 76 Hz) down to an average of 35 dB in the 4800- to 10,000-Hz range. A steady flow of automobiles traveling in multiple lanes can produce 90- to 100-dB sound in the upper frequency ranges where noise is most damaging and annoying.

The streets of New York City are famous for horn-blowing taxicab drivers and cross-town avenues constantly choked with delivery trucks. In that commercial nerve center of the nation, the flow of traffic never stops and the noise reverberates off glass and steel walls to reach far above street level, despite a nonenforced anti-horn-blowing law. A recent survey of traffic noise was made outside the sixteenth and seventeenth floor windows of three New York hotels. The audiometer showed the noise levels to reach 100 dB in the low frequency range, tapering down only to about 65 dB in the 4800- to 10,000-Hz bands.

. . . Control Efforts

One of the earliest attempts at police control of traffic noise is recorded in the *Times* of London of September 11, 1829. On that date, the *Times* reported, officers seized the horses of a stagecoach because the Lord Mayor wished to prevent disturbance to church services. As pointed out by Walter W. Soroka, professor of acoustical sciences at the University of California in Berkeley, increased production and improved transportation brought about by the industrial revolution resulted in expanded domestic and foreign trade and expanded competition. "The result," he said, "has been that cost and performance have ever since primarily determined the design of engineering systems, with noise often lagging far behind or totally disregarded, whether annoying or even deafening. Only recently has urban noise pollution received spasmodic, but increasing, attention on a more organized scale."

In 1922 the Public School Principals' Association of Newark, N.J. took the lead among cities of America requesting "noiseless" pavements in the vicinity of schools. Constant noise caused by heavy vehicles running on rough pavement kept pupils and teachers from working effectively. In 1924, the Joint Committee on the Noise Nuisance of the Civic Club of Philadelphia offered

the City Council a number of suggestions for decreasing unnecessary noise in the city, predicting "a vast increase in public
comfort" as a result. The Civic Club called for ordinances
diverting heavy traffic from residential streets at night and
making illegal unnecessary horn blowing, and enforcement of
a law forbidding street vendors from calling their wares. (Only
those of us who predate television can remember the startling
sound of a newsboy calling "extra" at two in the morning,
heralding some extraordinary turn of events in the affairs of the
world.)

Two years later preliminary plans were announced for a
country-wide effort to end "ear-splitting, nerve-wracking, sleep-
wrecking, health-destroying" noise. The movement was started
by the American Society of Safety Engineers, which hoped to
secure scientific cooperation from physicians, psychologists, and
psychiatrists to ascertain the harmful effects of noise, but the
effort apparently died amidst the general apathy of the public,
lawmakers, and industrialists.

First audiometric measures of New York City noise levels were
made in 1925-26, with a second survey in 1928 which extended
measurements to upper floors of buildings and subsurface sounds
in the subway. The surveys were made by Dr. E. E. Free, Science
Editor of *Forum*. The magazine then framed a model code by
which communities could levy fines against the owner of any
automobile, truck, streetcar, or other vehicle emitting unnecessary noise because of loose parts or bad adjustment. This model
code, which met general apathy, also prohibited wide gaps at
rail crossings and the blowing of automobile horns on streets
equipped with traffic lights. In January 1929, the *Engineering
News-Record* published an editorial stating that "we have al-
together too much noise for comfort and something ought to be
done about it. So far nothing has been done, because the dollars
and cents meaning of noise nuisances has not yet come to the
fore." The dollars and cents factor in the lack of control over
road noises, forty years later, remains the overriding reason why
the motor vehicle is permitted to assail public ears virtually
unchecked.

"Since the earliest recorded surveys of city noise," Professor
Soroka commented, "motor vehicle traffic has been shown con-

sistently to be the most serious source of widespread noise pollution. Moreover, this form of pollution has increasingly invaded suburban areas and the countryside as numbers of vehicles multiplied; intercity, interstate and transcontinental truck and bus traffic expanded; more and better road networks were built, and engine horsepowers as well as vehicle speeds also multiplied. Legislation to control this source of noise pollution has painfully developed over the decades, but still lags far behind in its effectiveness, both from the standpoint of prescribed limits and of enforcement." Early regulations to the effect that vehicle noise must not be "excessive or unusual" were vague, unenforceable, and often discriminatory in their interpretation.

A pioneering ordinance was passed in June 1954 by the city government of Milwaukee, Wisconsin, a law which for the first time spelled out permissible traffic noise in specific decibels. The regulation placed the limit at 95 dB measured not less than 20 feet from the right rear wheel of a vehicle in motion. The level was based on a jury determination of sound levels considered to be annoying. In the same year, Switzerland passed a nationwide law providing for a maximum of 90 dB at 24 feet for trucks and buses; 80 dB for autos; 90 dB for four-cycle motorcycles; 85 dB for two-cycle motorcycles, and 80 dB for motorized bicycles.

Other U.S. cities followed the Milwaukee pattern. In November 1958 Cincinnati, Ohio passed a revised ordinance specifying 95 dB as the permissible limit at 20 feet from the right rear wheel of a passing vehicle. The Beverly Hills, California law specified a maximum of 95 dB, measured not less than 5 feet from the noise source of the vehicle. Memphis, Tennessee specified 90 dB at 20 feet; Peoria, Illinois 85 dB at 50 feet. Averaging out the cities where experience with specific laws resulted in some enforcement success, the National Institute of Municipal Law Officers prepared a model noise abatement ordinance which provided for noise measurements at 20 feet from the right rear wheel of a passing vehicle, with 90 dB the maximum permissible level. It also called for state inspection stations for checking vehicles suspected of being too noisy.

In July 1965, the New York Legislature passed a state law which defines excessive noise for trucks at 88 dB measured 50 feet from the vehicle and prohibits noise beyond this level on its public highways. The noise is measured at roadside check stations where trucks must pass at speeds less than 75 mph. Judge Theodore R. Kupferman, New York Supreme Court judge and former congressman, commented upon the fact that truck manufacturers themselves helped the State Legislature to arrive at a reasonable noise limit for vehicles. Testimony at legislative hearings indicated that new mufflers available for diesel trucks and buses were capable of holding noise to 80 dB. However, since mufflers deteriorate with age and usage just as does any other mechanical equipment, the compromise figure of 88 dB was reached. "Now this I think is a responsible industry approach," Judge Kupferman said. "Maybe we can differ on the number, but at least they will recognize the problem and try to do something about it. I think it's something that will have to be done and recognized in other industries."

In California, which has been notably progressive in such dubious social advancements as freeway development, the Highway Patrol in January 1968 adopted a statewide vehicle noise regulation keyed not only to type of vehicle but relative speeds as well. In an open measuring site, vehicles accelerate in a straight line at maximum noise condition past a microphone stationed 50 feet away. The regulation prohibits the sale of any vehicle which exceeds the permissible limits which are:

1. Any motorcycle manufactured before 1970: 92 dB.
2. Any motorcycle manufactured after 1969: 88 dB.
3. Any motor vehicle with gross weight of 6000 pounds or more, manufactured after 1969: 88 dB.
4. Any other motor vehicle: 86 dB.

For vehicles traveling on the road, motorcycles and trucks above three-ton gross weight are limited to 88 dB at 35 miles an hour or 92 dB at higher speeds. All other motor vehicles should not exceed 82 dB at 35 mph or 86 dB at speeds above that figure.

The California Highway Patrol began limited enforcement of the new regulations in April 1969; but this state, with the finest of good intentions for protecting the public ear, may be used as a prime example of the extreme difficulties—approaching impossibility—of enforcing traffic noise limits. The difficulty begins with the question of providing adequate police manpower and the purchase of expensive sound-measuring equipment. Approximately 10 million vehicles travel California streets and highways. Policing each of these vehicles, both in-state and interstate, is nearly as complex as measuring the chemical content of each gallon of water flowing over Niagara Falls.

To detect violators of the code, the Patrol is using a microphone mounted on a tripod. This is connected by an extension cord to a sound level meter. The microphone is placed 50 feet from the center of the lane of a highway or freeway which is under surveillance. At the present time each of the six patrol zones in the state has a three-officer team attempting to enforce the law. Measurement is made of rattles, tire whine, and other vehicle noise as well as muffler and engine sounds. Each team, while it is engaged in the noise control effort, ties up three highway patrolmen and two patrol cars. In operation, officer number one *operates* the sound level meter, and officer number two *reads* the meter. Officer three parks his car in a safe location some distance away and maintains radio contact with the other two. If an offending vehicle passes the sound meter, officer three stops it and issues the operator a mechanical warning or a citation.

"It should be pointed out," states Lt. Bruce Emery, commander of California's Patrol Zone Five, "that the criteria as to where and under what conditions such surveillance can be made is highly restrictive. The area must be free of large reflective surfaces, signboards, buildings, fences, parked cars, etc. Furthermore, the surface should be free of water, with no grass or shrubbery over six inches high, and the road surface must be concrete or asphalt."

From that statement alone it is apparent that the California regulation, progressive as it may be, *excludes* almost all of the most critical areas where noise causes the greatest annoyance. All city freeway miles, highways, and streets are flanked by buildings, trees, signboards, or parked cars; therefore the regulation cannot be enforced in metropolitan centers. Houses and trees line suburban residential streets, so the measurements cannot be made there. The surface must be asphalt or concrete, so there can be no enforcement on any dirt or gravel road in the state. The areas remaining, which do comply with the regulation, are those miles of country highway and freeway where the noise problem actually affects the fewest people. Another enforcement complication, which should be obvious to the most simple-minded person alive, is the difficulty of sorting out the noise of an individual vehicle on a four- or six-lane freeway where hundreds of vehicles are passing any given point in any given minute. In Los Angeles rush-hour traffic, when four lanes of solid automobiles and trucks crawl along or are forced to stop with motors running, how could any sound-level meter or intelligent patrolman issue a citation to an individual motorist?

An individual motor vehicle is a point source of noise which creates a hemisphere of sound above and around it. When a line of vehicles follow each other closely, as on a busy express-way, the sound pattern resembles a half-cylinder extending outward and upward from the center of the traffic lane or high-way. A line of vehicles traveling close behind each other causes noise 9 dB higher than a single vehicle (in the case of trucks or motorcycles) and 15 dB in the case of automobiles.

In a nation then, which finds it so difficult to police the relatively small percentage of criminals in society, how can it be economically and practically possible to police such an elusive and transient condition as noise when the potential violators number 100 million? If any degree of effectiveness is to be obtained, approaches other than individual policing must be undertaken. Since the problem is sheer mass and numbers rather than individual aggressors, the possible solutions include achievement of (1) quieter vehicles, (2) fewer vehicles, and (3) relocation of fewer miles of roadway built perhaps of quieter materials.

Because diesel-powered trucks are the most severe noise source, state and federal governments should begin at this point to apply stringent enforcement of noise standards, which, themselves, should be made progressively more strict. Mufflers are available for trucks, but they do cause a power loss from the engine and their effectiveness deteriorates with time and use. The control of truck noise should be (except for the power of special interests) easier than policing automobiles because trucking companies already are controlled in other respects by state and federal authority.

The next step, after trucks and buses, would be the elimination of what are termed "deviant vehicles" by David Apps of General Motors Corporation. These include motorcycles, sports cars, and foreign cars. "These deviants are really deviant from an external noise standpoint," Apps said, "because there is no thought given to the design concerning noise. There are others who design purposely that way, and there are still others that start out very quiet, and the noise levels deteriorate, that is, increase because of improper maintenance." Apps also pointed out that the noise of tires on pavement probably adds as much to the sound level as the engines of passenger cars, and trucks as well. He said there are two basic types of tire design, one rib type which is relatively quiet, and a crossbar type which is noisy to begin with and grows noisier with wear. Apps speculated that most trucks and some automobiles are equipped with crossbar tires because they provide better mileage and thus are economically attractive. Requiring quiet tires on all vehicles might be one area in which enforcement could be effective.

. . . Future Possibilities

Quieter vehicles will be obtained only when a quieter and more efficient power source provides an economic successor to the diesel and gasoline engines. Both steam- and electric-powered vehicles offer hope for the future, even though both are opposed by the deeply entrenched petroleum industry, Detroit auto makers, and builders of highways. Steam and/or electric vehicles would offer a double benefit to our decaying environment because they are quieter and offer sharp reductions in air pollution levels. Major auto makers are conducting some desultory research in both steam and electric power sources, but the general attitude was best summed up in 1969 by General Motors President Edward N. Cole. "At the present state of technology the internal combustion engine is the best all around source for the automobile," he said. "We recognize that its emissions contribute to air pollution and we share the nation's concern over this important and complex problem, but the familiar growl of the internal combustion engine is, and probably will continue to be, the sound heard around the automobile world."

Cole gives no credit to the ingenuity which is axiomatic in American technology, nor do he and others admit that better vehicle power plants probably will not come to pass until forced by state and federal law. Many electric vehicles, produced by small and as yet unheard-of developers, already are appearing on U.S. and European streets. Though they do not produce the ego-building roar of 400-horsepower gasoline behemoths, they are ideal for delivery purposes and short city or surburban trips. William Lear, better known for aircraft developments, is one of a new breed which has turned attention to the steam car. In 1969 Lear exhibited a steam car at the International Automobile Show. It used a special chemical preparation in place of the water burners which characterized early American steam cars, and Lear promises that such engines can be built up to 500 horsepower. He has since turned his attention to gas turbines, but the Department of Health, Education and Welfare has evaluated design proposals for a 100-horsepower steam engine which could power a six-passenger car capable of 75-mph speeds. George Kittridge, chief of the research branch of HEW's Air Pollution Control Administration, said technical studies will be followed by research into financial feasibility. An operating vehicle was scheduled to begin tests in 1970. Weighing 3200 to 3600 pounds, it would be capable of accelerating to 60 mph in 15 seconds. It would be much quieter than gasoline-powered vehicles, and provide a bonus in reduced air pollution. Most experts predict it may be a decade or more before electric, steam, or other advanced power plants overcome the inertia built by 70 years of mass production of the internal combustion engine.

Aside from the intrinsic noisiness of autos and trucks, the location of highways is a major contributor to the spreading of high-intensity noise into the homes and neighborhoods of millions of people. The major blame lies with engineering which operates within the narrow confines of moving vast numbers of vehicles efficiently and safely, at lowest possible cost, without regard for the public welfare in the process. This is especially true of the multilane expressway which clashes across business, commercial, and residential communities. "Neither noise suppression laws or the electric car will remove the noise from the intense concentration of vehicles on the motorway," comments the British Professor Alan Proudlove. "The very elements of an urban motorway that make it efficient—traffic concentration, fronting buildings set back, high speeds—create a major source from which it is most difficult to prevent the noise spreading out in all directions. The zone of environmental intrusion of an urban motorway is far wider than is often realized—perhaps more than a mile, and certainly one-third of a mile."

Some states, including California, have taken community annoyance into consideration—after the freeways are in existence—and are trying to screen fumes and noise with flanking belts of trees and shrubbery. This helps, but probably is not adequate unless a quarter- or half-mile of dense trees could be provided on each side of the freeway. Economics prohibits this because of the high cost of right-of-way land. Other suggestions include bringing high-rise buildings up to the edge of the freeways so that the reverberating sound is thrown upward and not outward toward residential districts. Some engineers are toying with the idea of selling the air space *above* the freeways for the construction of massive, high-rise industrial, commercial, and apartment buildings. This concept leads toward the ideal—albeit an expensive one—of enclosing the freeways in a concrete tunnel, either above or below ground.

Reducing the number of motor vehicles is the most rational hope for reducing traffic noise, but this also will be most difficult to achieve within our industrialized society which worships the motor vehicle and all of the economic benefits produced by oil, road construction, auto production, and automobile insurance. Voices, including that of Rep. Jonathan Bingham, Democrat from New York, in recent years have begun sounding warnings that the automobile itself is threatening to throttle the great urban centers of the nation. "I think it has been amply demonstrated," Bingham said recently, "that the automobile is not effective transportation in urban areas and that building more highways is not going to solve the problem."

One basic solution is for America's major cities, probably with massive federal help, to provide modern rapid transit systems which will provide convenient, low-cost transportation to millions of people, taking their individual automobiles off the highway. The San Francisco Bay area at the present time is initiating the first new transit system installed in America in the past half-century. Others must certainly follow. Rubber-tired rail cars, capable of carrying 50 to 100 persons each, will hasten people from their homes to their places of work. One simple example of what this would do to the noise problem is that a single rail car (electric-powered and noiseless) could carry 50 to 100 people. The 50 to 100 cars thus left in the garage would represent total horsepower of 10,000 to 30,000 that is not heard in the multiple rumble of internal combustion engines. For the movement of freight, city planners already recognize that the limit has nearly been reached on the number of trucks which can roar into, and out of, metropolitan centers each day making deliveries of freight to stores and warehouses. There is no question that a radical new approach must be found to in-city freight handling, including the possibility that freight of the future may move in underground tubes, encapsulated in containers projected to their destination by air pressure and vacuum, much as the old-day change-making shuttles worked in department stores.

It is perhaps needless to repeat what many people are beginning to recognize, that new power plants must be found for autos and trucks, that local citizens must begin to shut down the indiscriminate juggernaut of highway-building, and that entire new approaches must be found for the movement of freight and people.

Then the surface world will become quieter.

CHAPTER 5

Intelligent zoning; quieter machinery for industry, mass transit, and the building trades, and the public's willingness to foot the bill can mute the roar of cities and soothe the jangled nerves of urbanites.

Sirens
and
Bells . . .

AT A BUSY intersection in the Ginza District of Tokyo, Japan is located a large illuminated sign which repeatedly flashes the noise levels in the immediate vicinity, just as American banks keep passing motorists and pedestrians informed of the time and temperature. Beneath the sign is a message stating that sound of 70-decibel intensity is the maximum safe level for a commercial area.

Unfortunately, the Tokyo sign has seldom registered a noise level below 88 decibels. In addition to traffic sounds, this apparently permanent built-in sound level is composed of the scrape and shuffle of millions of feet on pavement, the hum of millions of conversations, laughter and shouts, whistles, bells, sirens, pneumatic drills, and the multitude of other clattering objects and machines which mark the very loud heartbeat of large metropolitan centers.

"Tokyo is not only the biggest city in the world, it is also the noisiest," in the words of Minora Matsumoto, chief of the noise control section of Tokyo's city government. "There is no doubt that the high noise level creates serious physical, mental and spiritual problems for the people of this city." But Mr. Matsumoto shrugs his shoulders with the general helplessness of the situation. Tokyo has had a noise control ordinance for ten years, but it has no teeth and it has been unenforceable. More lately, the Japanese Diet passed a noise control law which enables local officials to levy jail sentences or other penalties against noise offenders. "The new noise control law can mitigate the problem to a certain extent," Matsumoto believes, "but there is no way the problem can be solved."

There is little doubt that Mr. Matsumoto's pessimism would be echoed around the world in those metropolitan centers progressive enough to have retained a noise control officer. To any such official, or governmental agency, the muffling of the busyness of millions of people must seem as hopeless as stopping a river's flow or damming the tide. The noise comes from every direction —heavy machines tearing down old buildings to make way for new, the sound of rivets linking massive steel beams, pneumatic drills ripping up pavement to make way for a new sewer line. All of these and more create the city's roar. In the country the quiet atmosphere is set into violent vibration by giant earth movers killing trees to make room for homes and highways, power lawn mowers, farm tractors which have replaced the quiet horse, and in America, the bark of millions of dogs which we seem able to feed better than most of the starving people in the world.

"Although the characteristics of traffic noise have been identified," the Committee on Environmental Quality wrote in 1968, "relatively little effort has been devoted to other sources of noise in the community. Serious consideration should be given to this aspect of the problem as well.

"Building construction and demolition, road building and repair installation and repair of underground utilities and services have increased at a phenomenal rate since World War II. The inconvenience these activities cause is largely unavoidable, but they also have been accompanied by the production of noise emerging from pneumatic drills, generators, air compressors, bulldozers, pile drivers, ditch diggers, concrete mixers and earth-moving and demolition equipment. Some of this, of course, cannot be helped, but some improvements can be made to reduce annoyance levels."

Since the beautiful islands of Hawaii became the fiftieth state of the Union, the song of the islands has gradually changed from the soft tones of ukulele and steel guitar to the island-jarring thump of pile drivers which awaken residents and visitors to Honolulu every morning to the sound of progress. Pilings must be driven to bed rock to provide a foundation for the new high-rise buildings which have changed the gentle sands of Waikiki Beach to a brasher version of the Miami Beach hotel row. Since the tourist trade is so vital to Hawaii's economy, it has been necessary, in fact, to prohibit pile driving from awakening the visitors at least until 8 a.m.

In addition to sounds of "progress" epitomized by bulldozer and air hammer, the sounds of industry often spill beyond the factory walls where workmen may be protected but the surrounding community is subjected to periodic high and low levels of pounding, thumping, grinding, and sawing. "The sources of occupational noise are also problems with which we must contend in the surrounding community," the Committee emphasized. "Although noise produced by these sources is likely to have less drastic effects on the housewife or her baby than on the factory or farm worker immediately associated with them, it is nerve-wracking and can reduce the desirability of property ownership in the area.

"Many of our annoying noise problems," the committee continued, "occur within the very areas in which we seek peace and quiet. Symbols of affluence such as go-carts, shopping centers and canned music contribute to our national wealth if not to community well-being. Metal garbage cans, lawn mowers, motor boats, portable radios and musical instruments may protect our

SIRENS AND BELLS

health, enhance the appearance of our homes, provide relaxation or educate our children, but they also may be annoying to others."

Abatement of these kinds of problems rests in a variety of approaches. Muffling power equipment and advancing technology can cure many of these problems. For others, greater respect for the comfort and well-being of neighbors, especially during the "quiet hours" is perhaps the key to greater progress.

"I am a hardliner about noise," stated W. H. Ferry, the noted conservationist. "I do not want to have to put up with the explosive exhausts of motorcycles, large and small. I have spent futile hours looking for a restaurant where my waffles and bacon need not be accompanied by the syrup of Muzak. Even in my own city of Santa Barbara (California), downtown merchants and city rulers have combined to bathe the passerby in an unending shower of that cozy sound known as background music. . . . The incessant tinkle of the Good Humor man has put me off street-purveyed ice cream for good. I feel anger on behalf of those thousands of human beings whose bedrooms abut freeways and aircraft runways. Outboard motors on remote mountain lakes, power mowers, riveters, bulldozers and their unholy kindred set my adrenal glands going with spasms of dislike.

"In all of these offenses against civilized life," Ferry declared, "I recognize not only technology, but bad technology—half-technology, if you like. For however well these instruments perform on their own terms, they perform badly on human terms. Their diseconomies outweigh their economies. So the few suggestions that I have to offer aim mainly at the elimination of the diseconomies; that is, at either perfecting the technology so that it will not be an insult to human sensibilities and the values of civilized life, or getting rid of it."

. . . Community Noise Surveys

The first comprehensive city-wide noise survey appears to be that undertaken by the New York Department of Health in 1929-30. It came to pass as a result of mounting complaints from citizens who found themselves increasingly disturbed by noise during sleep, recreation, or work. The New York Noise Abatement Commission was established in the fall of 1929 with representatives from medicine, neurology, otology, law, administration, acoustics, engineering, construction, and the automotive industries. Classification of noise sources—according to complaints of noisiness—revealed that traffic was responsible for 36 percent; transportation (such as trains and elevated railways), 16 percent; radios, 12 percent; collections and deliveries (including garbage and trash), 9 percent; whistles and bells, 8 percent; construction, 7½ percent; and miscellaneous (including human voice), about 7 to 10 percent. Twenty-five years later, in 1956, a newspaper poll of the most irksome noises in New York City placed refuse collectors and clanging garbage cans at the head of the list. Following this, in order of annoyance, were horn-blowing, truck and bus acceleration, blaring of radio and TV sets, aircraft noise, unmuffled exhaust, street repairs, sound trucks, construction riveting, and the whistles of hotel doormen trying to lure taxicabs. (It should be noted here again that for purposes of simplicity most of the noise survey references are based on the "A" decibel scale, which tends to depress the value of low-frequency sounds and thus more nearly approximates the hearing technique of the human ear. A number of attempts have been made to form statistical scales of *noisiness* and *annoyance* as measurable values, but these are highly subjective in nature and will be dealt with in somewhat more detail in the aircraft noise chapters to follow.)

Moved by the pioneering work performed by the Noise Abatement Commission in the early 1930s, the New York City Board of Transportation instituted a survey of actual noise conditions in the rapid transit lines, both subway and elevated railways. It was readily apparent that the elevated rail lines caused the greatest depreciation in adjacent property values, but equally apparent that relocating such lines underground was excessively costly. The study, however, did produce some results in defining what types of structures were noisiest and identified the steel wheel against rail as a prime cause of train noise. This study, and others which have been conducted in various metropolitan centers through the years, has recently come to assume new importance because of the necessity for planning and building

SIRENS AND BELLS

new, ultramodern mass transit systems as one means of preventing automobile and truck traffic from strangling our major
cities.

Periodic noise surveys have continued in New York, including
the newspaper poll of 1952. Another survey, conducted by the
New York *Journal-American* in 1959, was carried out at places
and times corresponding with the study done seven years earlier.
The sound levels were 5 to 11 dB lower in the new survey,
indicating that some progress had been made in controlling
New York's noise pollution.

In Germany, community noise surveys began as early as 1938,
and a recent noise map of Dortmund shows the noise levels of
important streets and industrial zones. London's survey, previously cited, showed that industrial noise predominated in 7
percent of the several hundred sample points, with construction
operations at 4 percent. A 1965-66 Tokyo noise survey provided
a map of noise contours, measured at various times during the
day and night, in seventeen typical areas in the city's residential,
commercial, limited industrial, and industrial zones.

On May 27, 1957, after seven years of study and development,
the Chicago City Council adopted an amended ordinance which
marked a major step forward in setting zoning requirements
limiting noise and vibration. Sound pressure levels were limited
to specified maximum values along residence district boundaries,
three classes of manufacturing zones, and business district
boundaries. Similar standards were proposed for Stony Point,
New York in 1954. Their limits were set on continuous nighttime
sound pressure levels at the boundary of the nearest residential
district, at the lot line, or twenty-five feet from the offending
establishment, depending on the nature of the districts involved.

The Warwick, Rhode Island performance code requires that
in residential or commercial districts, when measured at a
point twenty-five feet from an establishment where noise
originates (or at the lot line if it is closer), and in manufacturing
districts when measured at any point on the boundary of the
nearest residential district, the sound pressure level radiated
from a facility between 10 P.M. and 7 A.M. shall not exceed
specified limits in any frequency. In Oakland, California, the
code sets strict limits on nighttime noise spilling across property

lines. For daytime operations, it permits an increase of 5 dB in each frequency spectrum of sound.

Guiding principles for noise control proposed in Switzerland by the Federal Commission for Noise Abatement in 1963 set maximum levels in six zones: hospital and convalescent, quiet residential, mixed residential, commercial, industrial, and main arterial road. The nighttime basic permissible noise levels range from 35 dB in the hospital zone up to 65 dB in the highway zone. Allowable daytime levels are 10-15 dB higher. Recognition is given to zones in which special noise conditions occur, such as construction zones. Corrections upward also are permitted for intermittent noise, the frequency spectrum involved, and percentage of time the noise prevails. The city of São Paulo, Brazil, permits residential noise up to 45 dB at night, 60 dB by day; mixed zone: 65 dB at night, 80 dB during the day, and a range of 65-85 dB in industrial zones.

As indicated by the above listing of community noise surveys and the setting of rudimentary standards, proper city planning is a basic key to the control of noise while providing for the economic and personal welfare of a community's citizens and businessmen. A 1911 textbook on the city plan for Chicago states: "We recognize, for one thing, that city life is more intense and nerve-straining than life in the country. This means that our plan must aim to do away with unnecessary noises, smoke, dust, dirt, confusion. . . ." Urban noise control is an objective of both planning and zoning laws. Zoning invokes the municipal police power to protect the public interest through regulation of the use of land and buildings. Zoning is based on the principle that urban land may be divided into districts in which uses that tend to be compatible in their functional characteristics may be grouped so as to prevent or minimize conflicts among activities. Land uses thus are grouped or segregated in zoning districts according to their compatibility and regulated further as to function and form. The U.S. Supreme Court in 1926 took note of zoning as a means to "decrease noise and other conditions which produce nervous disorders." In 1954, the Court additionally acknowledged the protection of "public safety, public health, morality, peace and quiet, law and order" as traditional applications of the police power. It is the application of these

SIRENS AND BELLS

principles as part of municipal zoning which has led major cities to a gradual imposition of noise control limits upon segments of the community which may be regulated.

Increasingly careful attempts are made to quantify and regulate noise emissions with modern zoning codes, endeavoring to move beyond the basic full-spectrum decibel measuring stick to take various sound spectra into consideration. For example, the District of Columbia standards prohibit the spillover of noise which exceeds 39 dB in the high frequencies above 4000 Hz into residential and commercial zones. In the lower frequencies, the code permits a maximum of 79 dB.

While cities report some measure of success in regulating their internal noise, Dorn C. McGrath, director of the Division of Metropolitan Area Analysis for the U.S. Department of Housing and Urban Development, warns that local laws soon will be inadequate. "Maximum noise emission limits have been incorporated in zoning controls for land use in many American cities," Mr. McGrath said. "Coral Gables, Florida, where limits for even home air-conditioner noise are in effect, claims title as the quietest American city, as does Memphis, Tennessee, on a larger but not louder scale. In all such cities, people disturbed by loud noises from local sources are able to appeal to local officials and courts to obtain their relief.

"But maintaining the public peace and quiet in whole urban areas is becoming more and more difficult, even as the adoption of local noise-control ordinances spreads. This is because the contemporary problem of environmental noise is becoming metropolitan, rather than local, in scale. Today, the most serious noise intrusions in the urban environment are made by aircraft, highway traffic, and other transport systems that operate beyond the purview of conventional local controls."

Local havens of quiet in parks and homes are being invaded by more noise every day. It has been suggested that city noise in the United States is increasing at the rate of one decibel every year. Against an increasingly pervasive background of noise, every intrusion made by an occasional ambulance, lawn mower, or pneumatic drill is now felt more acutely by more people in urban communities. As a result, city noise has generated its own critical mass of controversy and a public demand for control at the appropriate level. Contemporary urban planning is increasingly concerned with development at the level of the metropolitan area, but this is the level at which noise control and planning for noise exposure suffers serious retardation.

. . . Source Control

Through zoning laws, cities and larger metropolitan areas now are experiencing some success in insuring beforehand that industrial noise will be properly buffered from residential and commercial areas. However, basic policing of noise laws still encounters the same difficulty as that previously mentioned in relation to road traffic; i.e., it is more difficult to police thousands or millions of point sound sources than to attack the source in itself, which in most cases is the manufacturer of noisy equipment. This fact relates back to Mr. Ferry's point that much of our noisy world is that way because of bad technology, or in other words, technology which has not taken noise into original design equations. For the manufacturers of most equipment, the primary task is to build a machine which does its job well and efficiently. Now the task is to maintain levels of economic efficiency while at the same time applying acoustic engineering at the design level. It has been shown in a number of instances that machines can be designed for quieter operation if the manufacturer is shown such requirements must be met. It follows that such requirements must be stated so long and loudly by the public that politicians and industrialists will listen.

Although subway and elevated railway noises had seemed to be melting into the nostalgic past of American cities, mass transit is making a strong comeback by necessity to reduce the number of motor vehicles in metropolitan areas. As explained by Professor Soroka of the University of California, "High speed rapid transit system development on a region-wide basis has had a resounding resurgence of local interest and of Federal support in a broad attack on the almost impossible congestion of workaday traffic by private automobiles between large urban centers and the surrounding smaller urban and suburban residential areas."

SIRENS AND BELLS

This has introduced a new dimension to the ground vibration and noise pollution problem, particularly for those residents situated in the vicinity of surface rights-of-way. The major source of noise and vibration is the random contact force between wheel and rail necessary to produce rapid acceleration, maintain speed, and provide the rapid deceleration of the train at stops. Although subways appear to be favored and provided in the larger centers capable of paying for them, in outlying districts surface and elevated structures are the rule. The new transit trains, however, need not be the noisy monsters of the past. Welded polished steel rails and aluminum-centered steel wheels with a polyurethane elastomer between inner aluminum hub and outer steel wheel promise to drop the radiated noise from 97 dB for a conventional steel wheel down to 90 dB for the new. Other quieting agents now available include elastomeric pads under the railroad tracks; sound-deadening materials applied to wheels, rails, and car underbodies; use of rubber tires instead of steel wheels, and use of vibration-dampening materials on box girder types of elevated bridge structures. Solid barriers alongside tracks also provide significant reduction in community noise. The benefits from most of these changes are just as great for the train passengers as for the residents who live near the tracks.

In the construction industry, the local noise pollution problem is intensified by pile drivers, jackhammers, poorly muffled air compressors at construction sites, concrete mixers, trench diggers, bulldozers, earthmovers, road scrapers, riveters, welders, electric saws, engine-driven chain saws, construction cranes, demolition equipment, and numerous other items of machinery found at construction sites. Noise levels for nearby residents range into the 90-dB levels.

According to Professor Soroka, a 1965 noise survey of earth-moving equipment at 16 construction sites covered scraper-loaders ranging from 360 to 1720 horsepower. Tractor-dozers ranged from 235 to 385 hp; compactors had 475-hp engines and graders 225 hp. The tests were run in connection with a study of damage risk to hearing on the part of equipment operators, and overall sound pressure levels of 90-120 dB were measured. (Most of the energy fell below the 600-Hz frequency band.) As for the earth-moving equipment, David C. Apps, of the General Motors Proving Ground, reports that the automotive industry has developed silencing packages for rear dump trucks, loaders, and other equipment. These silencing packages, he pointed out, are made available as "customer options." It is apparent that a building contractor probably would not exercise such expensive options until and unless he were required to do so by local regulation.

Silencing packages also have been developed for engine-powered air compressors. George Diehl, of the Ingersoll-Rand Corporation, reported to a Washington conference on noise abatement that his company has a persistent program underway for quieting compressors and pneumatic tools. He demonstrated a compressor which provides 900 cubic feet of air per minute and, without suppressing equipment, causes noise of 110 dB. By enclosing the compressor in fiber glass and increasing the ventilating fan size to keep the machine cool, the manufacturer was able to reduce the machine's noise output to 85 dB.

Another type of machine with extremely high noise output is the power-operated chain saw. One survey of chain saw operation showed sound levels of 110 dB, 117 dB, and 123 dB, depending upon muffler size, and whether or not a muffler was used. Such mufflers are available. The question is whether or not they are used. Professor Soroka reports that a "silent" pile driver has been developed both in America and in England. British Building Research Station tests have shown that jackhammer noise can be muffled considerably without impairing performance and may lead to legislation requiring the use of muffled jackhammers in towns and cities. Another new development is a type of vibrating plow with which one man can install underground residential telephone and power lines. The noise level is low compared with other ditch-digging apparatus.

It is encouraging that the nation's vehicle and tool manufacturers are being guided to some degree by an active Vehicle Sound Level Committee which is an arm of the Society of Automotive Engineers. The Society, in May 1966, approved a set of recommended standards for engine-powered equipment which would set a 90-dB maximum at a 50-foot distance for crawler tractors, dozers, loaders, power shovels, cranes, motor graders, off-highway trucks, ditchers, trenchers, scrapers, etc. The Society also has recommended that for powered equipment of 20 hp or less, the maximum permissible sound level be 72 dB at 50 feet. Thus, standards finally are being established for quieter lawn mowers, riding tractors, and residential snow removal equipment.

Mr. Apps advises that a major impetus for the provision of quieter machines—both road vehicles and other machinery—will be forthcoming when city, state, and federal governments not only purchase quiet equipment, but *require* quiet equipment as part of any contract signed for new construction. Civic-minded businesses, such as utility companies, could do the same.

"The public gets pretty much what it demands and fights for," Apps said. "Without concern of the average man on the street for these intrusions on his privacy, no leadership will emerge to better the situation.

"There are no great technical barriers to better control of vehicular and machine noise. There *is* a cost penalty, ultimately born by the citizen for such improvements. If he demands these, they will be forthcoming."

The citizen is beginning to be heard. And the technology, which caused the noise problems in the first place, will be called upon to bring more silence into our world.

SECTION II

THE COST OF HIGH-SPEED TRAVEL

CHAPTER 6

Government agencies pass the buck
on airport noise while millions
of people near airports are bombarded
with noise levels well above those
capable of causing permanent
hearing loss in industrial workers.

Jet
Thunder . . .

UNITED STATES PATENT No. 1,219,702—issued on May 20, 1917—is one which would enlist the interest, sympathy, and enthusiasm of millions of Americans today. The patent was granted to Mary Clarke, a British banker, and Demetrio Maggiore, an Italian engineer, and covered a device called "Apparatus for Dispelling Aeroplanes and the Like."

World War I was still in progress. Fokkers and Sopwith Camels were dogfighting above the bloody battlefields of France and Germany, and the Lafayette Escadrille was converting its small band of daredevil pilots into a romantic legend that would stimulate the excitement and admiration of small boys for generations. The airplane, at that time, was only beginning to grow beyond the novelty stage and the sound of one of these putter-putts was so rare around the average town or countryside that a cow-pasture landing was almost as great an event as a circus.

But apparently Mary Clarke and Demetrio Maggiore did not like airplanes even then. Their "apparatus for dispelling aeroplanes" consisted of a large upright cylinder into which gas was introduced and lighted to create an explosion. The combustion products would burst out of the top of the cylinder in a large, turbulent smoke ring. This, it was claimed by the inventers, would create such a violent disturbance that air currents would be affected, "hampering the flight of aircraft or upsetting same." The records do not indicate if the apparatus was actually built, or if it succeeded in dispelling airplanes, but in later years it has been known for a farmer to heave potatoes, in futile frustration, at military jet planes making their final approach to a neighboring Air Force base. Both the aeroplane dispeller and the farmer's potatoes are symbols of the growing segment of our population which suffers disturbance and annoyance as well as physical and psychological illness resulting from the increasing range and scope of high-intensity sound from the air.

It started, essentially, at Kitty Hawk when the internal combustion engine was wedded to propeller and wings. It reached its peak of romance in 1927 when Charles Lindbergh flew the Atlantic. Two years after that, America's outstanding rocket pioneer may have been the first man to attract the attention of police authorities with the noise of an aerial vehicle. This was Dr. Robert H. Goddard, who began experimenting with rockets as early as 1908. On July 17, 1929, he launched an experimental liquid-fueled rocket from a field near Auburn, Massachusetts. The 11½-foot-long rocket rose screaming 20 feet above the 60-foot launch tower, turned over, and crashed 171 feet away. While Dr. Goddard and his assistants were examining the wreckage, a police patrol car, two ambulances, and a convoy of autos with sirens and horns blowing roared up to investigate what was assumed to have been an explosive disaster. Although no one was hurt, Dr. Goddard eventually was convinced he should move his experiments, later supported by the Guggenheim Foundation, to more remote areas, such as the White Sands Proving Grounds in the New Mexico desert. Dr. Goddard died a quarter-century ago, but the rocket development he pioneered continues. In July 1969 the National Aeronautics and Space Administration sent America's first three lunar astronauts to the surface of the moon. These men ascended into space aboard a Saturn V rocket which in its main stage generates 7.5 million pounds of thrust and noise measured above 170 decibels—probably the loudest man-made noise short of the hydrogen bomb. The Saturn V sound wave, which vibrates heavy steel beams and buffets the human body with low-frequency sound at a distance of three and more miles, so far is heard only by those residents around Cape Canaveral, Florida (and the millions of visitors who come at launch time), but it epitomizes the immense power—and consequent noise—which modern technologists are capable of building into their flying machines.

The Saturn V rocket booster is nearly 2000 times more powerful than the largest commercial jet engine in operation today, but rockets and turbojets typify the new tools man is using to forge the wave of so-called progress. The headlong rush of technology and demand for speed in the spread of commerce, industry, and tourism around the world ushered in the jet transport age in the mid-1950s. Jets, like automobiles and the freeway system, are satisfactory for the work they were designed to do. But in the design and manufacture of jet engines, the creators thought only of an efficient machine, not one that could live at peace with the people of the world. Air transport of freight and passengers is expanding at an enormous rate, and this expansion leaves a rising backwash of noise to plague earthbound creatures, particularly those living near the nation's airports. Just as commercial aviation approaches the threshold of the sound barrier in speed of flight, the time has come to begin breaking the barrier of airborne sound on the ground. As summed up by the national Committee on Environmental Quality: "The rapid growth of the air transportation system has resulted in a wave of public reaction to aircraft noise on and near major

JET THUNDER

airports around the world. The problem also is becoming serious at many smaller airports.

"It can be characterized as one of conflict between two groups—those who benefit from air transportation services and people who live and work in communities near airports. The conflict exists because social and economic costs resulting from aircraft noise are imposed upon certain land users in the vicinity of airports who receive no direct benefits. It is important that this situation be rectified in an equitable manner consistent with the public welfare and the orderly development of air commerce."

The question of aircraft noise is extremely complex. In the early days, airports were located far from city centers for purposes of clear takeoff and landing safety. As commercial air transport progressed, most notably since the end of World War II, many secondary industries and commercial centers took advantage of the high volumes of business the airports generated. Then home developers (and city zoning boards) permitted thousands of homes to be built around the airports. Finally a situation exists in which a city's airport is hemmed in on all sides by people who moved there *after* the airport was built. At the same time, the noise of large jet transports spreads over a wider and wider area while the industry requires expanded ground facilities for operational growth. The situation, again, is a conflict between a vital national and local asset on the one hand and the quality of individual life on the other. This dilemma also illustrates how political sloth and apathy have allowed the noise of aircraft to reach epidemic proportions before appropriate governmental action takes place.

Aside from haphazard governmental control in the development of airport communities, the aviation industry itself is to blame for placing both airport operators and airline managers in an intolerable dilemma. People irritated by noise complain not to the manufacturers who built the noisy machines, but to city officials and airline operators.

"The aviation industry ushered in the jet age with a sort of 'public be damned' or, at least, 'the public can't stand in the way of progress' attitude," states Austin B. Brough, president of the American Association of Airport Executives.

"Subsequent events have proven otherwise. An appreciable sector of the public has not learned to live with the noise and it is quite evident that it is not about to learn.

"To date the instigators of the noise—the manufacturers—and the perpetrators of the noise—the operators—have been able to walk away from much of the turmoil caused by excessive aircraft noise," Brough continued, "leaving the airport the unenviable job of trying to operate and improve a facility to accommodate jet aircraft while attempting to placate an unsympathetic neighborhood. This cavalier attitude on the part of the aviation industry is backfiring."

It is a pity, but perhaps typical of America's free enterprise system, that a situation must backfire in the form of legal action or governmental control before any meaningful action is taken, but this is the case with aircraft noise. Since the jet transport plane became a fact of American (and world) life fifteen years ago, local, state, and federal government agencies have continued to pass the noise buck back and forth without concerted and coordinated action. As the decade of the 1970s opens, some appropriate action is being taken but only after politicians have been subjected to the collective wrath of the people, who are beginning to take their case to court.

Aviation Growth

Since introduction of the large jet transport, commercial aviation has grown far faster than anyone had predicted. This is a special reason why the noise problem has gotten out of hand so quickly. A number of fledgling airlines were operating as early as 1926 with a motley variety of single-engine open-cockpit planes, and United Air Lines was formed by combining three others. After Lindbergh's flight to Paris, aviation excitement spurted, but commercial airlines nearly died in 1934 when President Roosevelt canceled contracts and ordered the Army Air Corps to carry the mail. Out of the depression of the 1930s came a sturdy twin-engine aircraft, the DC-3, which became the workhorse of the airlines and was a major factor in moving military men and freight throughout World War II. As for routine airline travel, immediately before World War II only 3 percent of intercity passenger traffic was by air in the United States. After the war—which spurred development of four-engine prop planes, the turboprop engine, and the first rudimentary military jets—air travel jumped to 17 percent of the total. The increase was rapid in succeeding years; but the greatest upsurge was due to introduction of the four-engine jet transport, principally the Boeing 707, which was born out of the military B-47 and B-52 bombers.

In little more than 50 years, the airplane had increased in speed from zero to 600 miles an hour, which cut a transcontinental journey to about 4.5 hours. In 1966, 110 million passengers traveled by scheduled airlines in this country, 90 percent by jet. That total was 16 percent higher than in 1965 and double the number of passengers who flew in 1958, the last year before jet service began. Around the world, in 1968, airlines (excluding Russia and China) carried 261 million passengers a total of 191.5 billion passenger miles. This was a 12 percent increase over 1967. Air freight totaled 5.4 billion ton-miles, 29 percent more than the year before. The global air traffic total, in the 116 countries which are members of the International Civil Aviation Organization, in 1968 was estimated to be 3.4 times greater than in 1959.

Although jet airliners and their operators receive the major blame for today's revolutionary growth in aircraft noise, it is not so much because of numbers as the fact that their operation focuses upon densely populated areas around major airports. For example, in 1962 there were only about 2000 airliners operating in the United States, compared with 25,000 military and 75,000 private and business planes. It should be remembered that the latter, although many use isolated small airports, also contribute to the overall sound level of major cities such as Chicago, New York, Washington, and Los Angeles. The question then is not so much the *numbers* of commercial jetliners in service (a point which will be explored additionally) but the number of times during the day and night that all types of aircraft land, take off, or run up engines in flight test facilities. The Federal Aviation Agency reports that in 1969, O'Hare field at Chicago was America's busiest airport with 690,810 takeoffs and landings. Second was Los Angeles International Airport with 594,486. In third place, oddly enough, was Van Nuys, California (567,973 plane movements during the year), which does not accommodate the large commercial jets on a regular basis. Chicago thus experienced nearly 1900 aircraft takeoffs or landings per day, and Los Angeles had a hefty average of 1650. That averages more than one aircraft movement per minute through a 24-hour period.

As for the growth of scheduled airlines around the world, there were 3144 jets in operation (excluding Russia's Aeroflot) during May 1969. About 950 more were on order for delivery in the early 1970s. In the United States scheduled airlines expect to emplane 3.5 billion passengers during the next decade, compared with 1.2 billion during the 1960s.

Alan S. Boyd, shortly before he left office in 1969 as U.S. Secretary of Transportation, discussed the explosive growth of commercial air transport in a speech before the Society of Experimental Test Pilots. "Our system of transportation represents an investment of $500 billion," he said. "It meets the needs, with varying degrees of effectiveness, of 200 million people. It accounts for one of every six dollars in the economy; provides jobs for nine million people, and unites a continent.

"Yet the increasing demands on this system already strain its capacity in some areas, and the growth to come—compounded by concentration of that growth—could bring it near collapse." By 1975, he remarked, the number of private aircraft will have doubled and commercial air travel will have tripled.

The growth in transportation demand is exerting its greatest pressures today in aviation. U.S. airlines carried 70 million passengers in 1963 and 150 million in 1968. Five years ago everyone predicted a 28 percent growth for the period. It was actually 114 percent. The number of scheduled passengers is doubling every five years, and by 1977 one million people will board commercial airliners every day. General aviation will grow even faster—from 100,000 aircraft today to an estimated 150,000 by 1973. Traffic control centers, which last year handled 15 million flights, will have to manage 30 million within five years.

This rate of growth severely challenges the federal government as operator of the airway system, and state and local governments as operators of the airports. The most serious problem inherent in any technology on the horizon, is noise.

In 1938, U.S. airlines, domestic and international, had operating revenues of $58 million. In 1968, the business had grown to $7.2 billion. The air transport industry thus is unquestionably of great value to the economy and commerce of all nations. Yet this value is enjoyed directly by only a relatively small percentage of the people. The rapid expansion of both passenger and freight movement by air often is taken for granted, but today the expansion of air travel is threatened by the very modern jet equipment which was designed shortsightedly without taking the public sensitivity to noise into consideration.

"Airport and community noise make it extremely difficult to get a community to accept either another airport or expansion of an existing airport," commented Secor D. Browne, assistant U.S. Secretary of Transportation. "The community wants an airport handy the day it wants to travel, but somewhere else the day it doesn't want to."

What the community really wants are quiet aircraft. Somehow a balance must be found between the tremendous value of the air transport industry and the harm which jet aircraft now cause to millions of ears and private property surrounding major airports.

... Airport Noise Levels

People living in the vicinity of an airport, either civil or military, are exposed to noise under three conditions: when a plane is taking off, when it is landing, and when the engines are being run up for ground test. During a jet takeoff, the noise is intense but much of the sound is in the lower frequencies. Quite another sound is heard during the descent of a jet liner. Although the thrust level and low-frequency sounds are low, the high-frequency sound from engine turbines is loud and penetrating. The full-spectrum sound pressure near a jet plane at takeoff may range from 130 to 140 decibels, sufficient to cause auditory pain. If a person were exposed constantly to such sound levels, he would be deafened permanently in a relatively short period of time. This is why all ground-based personnel wear carefully made earmuffs when they are working around a jet transport with engines operating.

Different types of aircraft produce different noise spectra. The sound from a military jet fighter at takeoff power extends over the audible frequency spectrum, but a low-frequency roar is dominant. Some civil jet transports use similar engines, but in-flight sound suppressors alter the frequency spectrum. Their sound at takeoff is richer in high frequencies than a military jet engine. During idling or taxiing the predominant sound from jets is a high-pitched whine made up of many pure tones emanating from the engine compressors. The sound from a piston-engine or turboprop aircraft is predominantly low frequency, largely attributable to the propellers. Helicopter sound is similar.

The loudness or noisiness of a sound as it is *perceived* by people depends upon the predominant frequencies and the range of frequencies present. For purposes of simplicity throughout this book, most sound-level figures have been presented in sound pressure level on the "A" weighing scale, which suppresses low-frequency noise as does the human ear. However, for purposes of measuring aircraft noise, laboratory researchers have attempted to correlate sound power and sound pressure levels with subjective rating methods which would provide a scale of noisiness at the listener's ear. One resulting scale, and that which is in most common use in discussing aircraft noise, is perceived noise in decibels, or PNdB. The basic point to remember about this scale is the way in which it differs from sound pressure expressed in dB. On the sound pressure scale, which we have mentioned before, a 6-dB difference in decibel reading means a doubling or a halving of the sound. On the PNdB scale, generally, an increase of 5 PNdB in a sound is a 50 percent increase in its "noisiness," as judged by the listener, and an increase of 10 PNdB represents a 100 percent increase, or doubling, of noisiness. For example, a reading of 105 PNdB is 50 percent less noisy than 110 PNdB. A sound of 115 PNdB is twice as noisy, or double, the 105 PNdB.

The PNdB scale is especially valuable because it helps to dispell frustrating problems which acoustic engineers and researchers encounter in attempting to translate sound pressure and sound power levels directly into measures of noisiness. For a given physical intensity of sound, people (the listeners) judge the jet plane sound as noisier than the propeller sound, since the jet sound has more energy in mid and high frequencies. Let us take for example a propeller plane takeoff, a commercial jet takeoff, and a jet landing, each of which generates equal sound pressure level of 100 dB. On the perceived noise level scale (PNdB) the propeller plane takeoff rates 102, the jet takeoff 111, and the jet landing 114.

The difference has to do primarily with the different frequencies and added annoyance caused by higher-frequency noise. Frequency spectrum information also is important for determining the reduction in sound with distance. This assumes importance in relation to aircraft noise. In normal outdoor conditions, the noise level decreases as it spreads outward from a source at a rate of 6 dB for each doubling of distance. Thus, if the sound pressure level were 112 dB at 100 feet from a source, it would be 106 dB at 200 feet, 100 dB at 400 feet, 94 dB at 800 feet, etc. This represents the reduction in sound which occurs with distance just as light decreases with distance from a radiating light bulb. However, there is an additional decrease in sound due to air absorption in the middle and high sound frequencies. This reduction increases with frequency, and above 5000 Hz this form of attenuation can exceed 10 decibels per 1000 feet of distance.

In reducing these complexities (additionally complicated by the listener's position relative to the aircraft, air temperature, etc.) to *perceived* noise levels, we find that a four-engine propeller airliner on takeoff generates about 115 PNdB at a distance of 400 feet, tapering down to about 80 PNdB at 10,000 feet. Jets are louder, both in terms of greater power and higher-frequency sound. A standard four-jet airliner generates 136 PNdB at a distance of 200 feet, with the sound dropping sharply as it moves outward and upward. At 10,000 feet, the PNdB is about 85. The modern turbofan jet engines, used for all new aircraft, provide about a 50 percent, or 5 PNdB, noise reduction at comparable distances. On landing approach, both the standard and turbofan jets provide about equal perceived noise-level contours. These range from 80 PNdB at 4000 feet to 127 PNdB at 200 feet. It should be noted throughout that distances from the sound source apply both vertically and horizontally and a combination of both.

From these figures, it is apparent that many millions of city dwellers are exposed to intermittent noise levels well above the 85 to 90 dB determined capable of causing permanent threshold shift in industrial workers. The major difference, as it relates to hearing, is that the aircraft noise is not constant. However, it *is* approaching constancy at the major airports such as Kennedy at New York, O'Hare at Chicago, and Los Angeles International. There are a number of other variables, also, which make it difficult to determine how much sound is suffered by individuals living in certain locations. Wind, for example, will carry the aircraft noise toward some, and away from others. The loudest sound from a jet occurs to the side and at an angle behind the wing; thus most severe noise occurs after the plane has passed the observer. Takeoff and landing techniques also have a bearing upon the PNdB on the ground. A steep angle of climb or descent greatly decreases the area which is disturbed by noise. A lightly loaded plane will climb to high altitude quickly, but a heavily loaded one will labor along at lower altitude for a longer period of time, thus spreading its curtain of noise over a far greater ground area. Communities surrounding high-altitude airports, such as mile-high Denver, Colorado, have higher noise exposure than sea level communities because planes require more runway distance and greater airspeed to fly in the thinner atmosphere.

There are three basic approaches to control of aircraft noise: (1) designing aircraft and power plants to minimize noise; (2) establishing operational procedures to keep noisy aircraft as far as possible from concentrations of people, and (3) planning, zoning, and redeveloping communities for compatibility with airport operations. These will be discussed in the following chapter, but before any of the three may be accomplished in unison with the others it is necessary to establish what sound levels are permissible at what distance from an airport, then establish appropriate rules and procedures which can be enforced.

... Standards and Regulations

Until most recent years, local, state, and federal authorities have hidden their heads in the sand (and that *is* one way to avoid hearing) on the question of aircraft noise. Local and state officials, essentially lacking the power to control an interstate operation, watched the noise problem grow while wrestling with the nebulous problem of community criticism. The federal government seemed to rely on the vague hope that communities would adjust to the rapid increase in aerial traffic or that perhaps the problem would just go away. It didn't, and it won't. Now the federal government is attempting to impose controls upon a situation which in some areas is virtually beyond control. The reason it is virtually uncontrollable is that the pattern of air travel already has built to the present level where the public finally is rebelling against the noise. Again, the aviation industry could have saved a great deal of misery for government agencies and airline operators if it had policed itself instead of waiting for inevitable government intervention.

A number of other nations have been more enlightened and willing to act than the United States. Some cities in the United States and elsewhere have attacked the noise problem without waiting for specific standards by requiring change in flight patterns to avoid densely populated areas. Others have established curfews prohibiting jet movements or engine runup between midnight and 6 A.M. or 11 P.M. to 7 A.M. Such cities are Sydney, London, Tokyo, and Paris.

"This simply means," stated Secor Browne, "that you cut down the time an aircraft can work; therefore, the cost of the system goes up. An airport curfew is particularly painful when you deal in the field of air freight. People by and large are pretty unenthusiastic about traveling, say after 10 or 11 o'clock at night or before 7 or 8 o'clock in the morning. Freight doesn't care. Late hours are prime freight shipping times."

Controlling the pattern of flight operations has been the principal method of attack at Heathrow, the largest airport serving London. There a number of monitoring points were set up in a pattern around the field and the ground noise level is controlled at limits of 110 PNdB by day and 102 PNdB at night. Aircraft on takeoff are required to rise steeply so as to be at least 1000 feet high when they are over residential areas and are further required to maintain reduced thrust until reaching 3000 feet in altitude. Mark Colbeck, division controller for civil aviation, commented: "Aircraft noise is an unpleasant nuisance for many people and causes real distress to some. It can be damaging to amenity, making less tolerable the environment in which we live.

"The noise abatement measures have certainly made the annoyance much less than it would otherwise have been. The operation of modern jet aircraft at major international airports raises a host of complexities and noise abatement controls must take account of the over-riding needs of safety of people in the air and on the ground. We are constantly seeking to improve the ways in which the major social problem is dealt with."

Frankfurt-Main, West Germany's largest airport, established a noise abatement commission three years ago, and the Stuttgart airport was one of the first in the world to use a centralized computer system to analyze noise readings from many monitoring points, with instantaneous reports permitting more careful control of aircraft landings and departures.

In November 1969, the U.S. Federal Aviation Administration, after much prodding by Congress to act under Public Law 90-411, announced noise limits for areas surrounding American airports. The rules, effective December 1, 1969, require all new classes of commercial aircraft to meet the prescribed limits, which range up to 108 PNdB on approach, takeoff, and along runway sidelines. For *new* aircraft, this would average about 10 PNdB reduction from the noise of standard three- and four-engine jet transports.

Keyed to the weight of aircraft, the new rules permit noise on approach up to 102 PNdB for craft weighing up to 75,000 pounds, and up to 108 PNdB for aircraft weighing 600,000 pounds and over. (The noise is measured one mile from point of touchdown.) Sideline noise limits also range from 102 to 108 PNdB, measured from .25 to .35 of a mile from the runway centerline. On takeoff, the noise limits range from 93 to 108

PNdB, measured 3.5 nautical miles from the start of the takeoff roll.

The FAA, in announcing the new regulations, made it clear that the long-term objective is to reduce noise around municipal airports to 80 PNdB, but it would appear the public will have a long time to wait for this. In the first place, the new regulations permit tradeoffs so that if an aircraft is quieter than the limit at any one of the three measuring points, it would be permitted to be proportionately noisier at one of the other two. In the second place, the rules do not apply to the older standard jets now flying.

The FAA is considering steps by which the older jets would be required to install sound-suppressing material to meet the regulations. There is no indication when this retrofit rule will be applied because it is a very costly process. In the meantime the older jets will be permitted to make just as much noise as they have in the past. Secor Browne estimates that retrofitting older aircraft for noise suppression will cost from $1 million to $2 million per four-engine jet. This burden will be placed upon the airline operators and in Browne's words, "It's got to come out of something. It comes eventually out of the ticket, out of the waybill, out of the freight charges."

"As often happens," stated Austin Brough, of the Airport Executives Association, "the regulations represent a compromise with which no one is particularly happy. The requirements will accomplish one thing—they will stop the further escalation of noise around airports.

"On the other hand, they will do little to correct the present unacceptable noise situation for many years.

"It appears that it is just a matter of time, and continued public pressure, before the aircraft operators will be forced into a retrofit for noise reduction program. Would it not make sense to take the initiative rather than being forced to act?"

That question, in essence, echoes back through the years since the first aircraft engine was manufactured.

NOISE QUIZ

(1) What president of the U.S. challenged all levels of government and the universities to come to grips with the problem of noise in the everyday government?

☐ Abraham Lincoln   ☐ FDR
☐ JFK   ☐ LBJ

ANSWER: LBJ

(2) Last year the Surgeon General of the United States reported that between six and sixteen million American adult males were being made deaf. What was the cause?

☐ Occupational noise
☐ Household appliances
☐ Rock'n roll music

ANSWER: Occupational noise

(3) What is the weakest sound that the healthy human ear can detect?

☐ One decibel   ☐ Ten decibels

ANSWER: One decibel. The decibel is the basic measurement for a scale of noise levels.

(4) How many decibels are the loudest sounds experienced in the everyday environment?

ANSWERS: Subways and electric blenders can exceed 90 decibels. Loud power mowers can exceed 100 decibels. Unmuffled motorcycles and construction noise can exceed 110 decibels at 10 feet. A jet take-off at 200 feet exceeds 120 decibels. A 50 HP siren at 100 feet exceeds 130 decibels. Noise above 85 decibels is dangerous to hearing.

(5) Building Codes can contain sound-insulation provisions. The first such building code appeared in Europe around 1938. When did the first American municipality adopt a building code with noise control provisions?

☐ 1940   ☐ 1950   ☐ 1960
☐ 1965   ☐ 1968

ANSWER: 1968. New York became the first American city with a building code with minimum provisions for making multiple dwellings less of an acoustical slum.

It was signed into law by Mayor Lindsay November 6, 1968.

From *Quiet, Vol. 1, No. 1, Spring 1969. Published by Citizens for a Quieter City, Inc., New York, N.Y. 10022. Prepared by Robert Alex Baron. Reprinted by permission.*

CHAPTER 7

*Jumbo jets may help the designers
of quieter airplane engines win their
nip-and-tuck race with the phenomenal
growth in air traffic that cancels out
their improvements.*

New
Aerial
Giants . . .

LOS ANGELES, CALIFORNIA, is a prime example of a city in which the aircraft noise problem has mushroomed virtually out of control despite actions which once appeared to place this West Coast community at the forefront of airport planning. In the eleven years since jet transports went into operation, business at L.A. International has doubled and redoubled with plans now progressing for a complete net of new regional airports to augment the major field.

The Los Angeles plan, and land purchases, had allowed for

growth. For the first nine years of the jet age, the airport utilized only runways south of the main terminal, but it was always known that a north runway must be put into operation at some later date. In 1968 airport authorities found it necessary to begin using the north runway to handle the overflow of aircraft movements. In the intervening years, however, neighboring community planners and school boards had gone blithely along, permitting major residential districts and schools to grow up in the vicinity of the north runway. The result: a swarm of enraged citizens protesting the noise, and carrying their case to city council chambers, the state legislature, and the courts.

By February 1969 the Los Angeles Department of Airports had been named in damage suits and claims totaling nearly $5 billion, including a $2.87 billion lawsuit filed by the Inglewood Residents Protective Association. This demanded $10,000 for each resident and $15,000 for each property owner. The emotional heat was expressed vividly by attorney Alvin G. Greenwald, representing the Inglewood complainants. "Residents," he declared, "won't tolerate being the garbage can for the noise, smog and other pollutions from the airport. Just as is true of smog, aircraft noise is a social cancer, and this cancer, not society, must be removed."

In November 1968, the neighboring Westchester school was closed because of noise interference with teaching and the 550 elementary pupils moved to another school. The school board later announced a second school would be closed by 1970. Officials estimate that seven schools in Westchester will require soundproofing to permit continued education, and twenty additional schools lie under the main approach pattern to the field. The Los Angeles Times, in an editorial, commented, "The jet noise problem raises a perplexing question: should one government agency accept fiscal responsibility for the damage it causes another? We think it should."

Clifford A. Moore, general manager of the Los Angeles Department of Airports, explained to the California State Assembly that jet noise control has been "completely pre-empted" by the federal government. He also pointed out that pending damage claims, if they were granted by the courts, could close down the Los Angeles Airport System and "that would bring economic disaster to southern California." Preliminary court decisions were conflicting and confusing. In one suit brought by property owners, Superior Court Judge Robert W. Kenny ruled that the city and nine airlines must file answer to injunctions brought against them. He held that property owners have the right to urge that airport jet plane traffic be handled without excessive noise and suggested a remedy might be compelling the government to pay for property no longer usable by the owner because of the government's activity. In another superior court considering a separate property owners' lawsuit, Judge William H. Levitt ruled that "the perceived noise level produced at their properties by air traffic and ground operations at the airport was normal for such an area, between 60 and 95 PNdb." He said inconvenience and annoyance "is accepted today as one of the inevitable burdens of modern living near metropolitan airports." Testimony showed that property values in the area had actually increased because the desire of people to live near their work apparently outweighed the annoyance. Judge Levitt held that the "nuisance value of jet noise alone cannot be held compensable without turning the construction and operation of airports into an intolerable public burden."

However, a completely opposite decision was handed down on February 5, 1970 by Los Angeles Superior Court Judge Bernard S. Jefferson. In what promises to be a landmark decision, Judge Jefferson awarded $740,000 in noise damages to 539 owners of property who live under the main approach pattern to the city's International Airport. The damages, ranging from $400 to $6000 for each piece of property, are assessed against the city for depriving the individuals of full enjoyment of their property.

"There is a significant difference," Judge Jefferson wrote in his decision, "between the noise emanating from jet aircraft and that coming from automobiles and trucks on a street or freeway. The difference is so pronounced that the legal consequences of jet noise should not be the same as the legal consequences of street and freeway noises of cars and trucks." He added that the high frequency of jet noise bothers the human ear and under extreme conditions prevents conversation and interferes with radio and television reception.

So here and there across the nation, the legal basis upon which excessive noise may be compensable as civil damages upon the suffering public at large is just beginning to be formed. In the meantime, government, airline, and aircraft manufacturing officials still are floundering in a no-man's-land of trying to place blame for the noise of the jets. The blame, it appears, soon will carry a high price tag in dollars.

Speaking before the Southern California Aviation Council in July 1969, John Shaffer, Federal Aviation Administrator, commented that lack of adequate planning is the major fault. "Hemmed in by the city, plagued by traffic-jammed accessways and besieged by citizens groups complaining of noise," he said, "the average jet airport today faces the dilemma of having to expand with nowhere to grow."

Thus the example of Los Angeles is being, or will be, repeated throughout major metropolitan centers wherever the main jet air lanes go. Secor D. Browne said people "will have to learn to accept aircraft noise to some degree. A quiet airport today is one where there are no airplanes. The price today for a quiet airport is a serious setback in our transportation capabilities and in the local economies of our nation." Although the economics of transportation is important, the air transport system may well be forced to pause and take a breath while the rights of the average public are being honored.

This same theme was stressed by the Committee on Environmental Quality of the Federal Council for Science and Technology. "The most rational approach to resolving the conflict caused by aircraft noise," the Committee found, "is to reduce the adverse effect of noise to the lowest practicable level with an equitable allocation of costs."

From an economic viewpoint, airport noise is an "external" cost. This exists when the production of a commodity or service by one economic unit imposes an unfavorable or unwanted effect on another entity, for which payment is not provided. If some of the external costs of airport noise are not to be borne by the industry, then air transportation has an advantage over some other industries. The unfairness of this situation leads to the conclusion that the services provided by the air transportation industry should include external costs. *This means imposing the costs of alleviating airport noise on the industry and ultimately on the public.*

The question of how the public benefits from air transportation is important. One of the arguments raised against charging air transport users for noise reduction is that the air transport industry provides many economic and other benefits to the entire nation. Such widespread benefits, it is claimed, justify the use of federal funds to pay at least a portion of airport noise costs. There is little question that air transportation benefits the majority of the population of the United States, whether or not they have ever flown. A modern air transportation system is a national requirement, but its overall net value has not yet been assessed. The answer is not clear unless we know the value obtained as well as the value sacrificed to obtain it.

Noise Reduction

The current public outrage caused by jet noise comes as no surprise to the aircraft industry itself, or specifically the companies which manufacture jet engines. Ever since the first such engines were developed in the late 1940s, companies have been attempting to cope with the community relations problems caused by engine noise. When a test jet rig was revved up in the night, company switchboards were swamped with complaining calls, and through the years millions of public relations dollars have been spent trying to convince the public that the jet scream is "the sound of progress" or "the sound of the future." This line was swallowed to some degree as long as jets were limited to military aircraft which could be justified in the name of national defense, but that justification has quickly crumbled in the last fifteen years of jet air transport. The public asks quite pertinently why it must suffer for the convenience and economic benefit of a relative few.

"The distressing fact about aircraft noise," states Matthias E. Lukens, president of the Airport Operators Council International, "is that while the problem has been with us for so many years, even before the advent of jets, precious little has been done by the industry to provide substantial relief of noise.

"Leaders in aircraft manufacturing, airlines and government have failed to acknowledge the severity of the problem. Too few have accomplished too little. There's been a lot of talk, but no really effective action. And throughout these years of temporizing with the noise problem, new larger, faster and, in most cases, noisier aircraft have been produced at a phenomenal pace.

"But the industry forecasts that predict a tripling of passengers in 10 years ignore the problem of how we will provide terminal and runway capacity to accommodate such growth in face of the public revolt against aircraft noise and the aircraft which cause it."

It is all too obvious that the aircraft industry has concentrated its major effort upon creating more *efficient* machines, and quieting a jet *does* cause weight and efficiency penalties. But in fairness, it should be mentioned that the industry spent $50 million to develop in-flight sound suppressors for jet power plants before commercial jets went into service. By 1965 the air transport industry had invested $150 million more on the same problem. Later the early engines were replaced by the new turbofan engines (at a cost of about $1 billion) which improved operating efficiency and gave sound reduction of about 5 PNdB, or 50 percent.

The pattern, however, is clear. The sound suppression efforts have been far below the scale of growth in the air transport industry. During the time required for each new increment of sound improvement to be developed, the number of aircraft takeoffs and landings at major airports increased so rapidly that the noise improvement was cancelled out completely.

Today, however, the world is entering a new age of jumbo jets which offer promise of a new dimension both in operating efficiency and noise reduction. The first of these is the Boeing 747, which went into service in 1970 with a passenger and freight capacity more than twice as great as jet transports now in general service. Although the 747's four engines are more than twice as powerful as earlier engines, they are 5 to 8 PNdB quieter.

Following close behind the 747 are the DC-10 and the Lockheed 1011 TriStar, expected to enter service in the 1972 time period. These large planes will ease the noise problem in two ways. First, they represent the first generation of aircraft in which noise reduction was built into the engines during initial design. Second, with their large passenger- and freight-carrying capacity, the jumbos may reduce the number of airport landings and takeoffs by a factor of two or three. This advantage of fewer residential flyovers will be a temporary benefit gradually decreasing as air travel expands, but the respite so provided will give another three to five years for even quieter aircraft and engines to be developed. It may well be that the years will catch up with us faster than engine improvements because, in addition to the large anticipated passenger increase, James Montgomery, Pan American Airways' vice president of sales, predicts the volume of air freight by 1978 will be nine times greater than today.

Some progress is being made toward development of what the aerospace industry terms the "quiet" engine, although a truly quiet jet engine is still far in the future. As R. Boyd Ferris of the International Air Transport Association commented recently, "Where noise is concerned, there are few, if any, major battles to be won. Success will hinge on many small victories."

Three of the major jet engine manufacturers are Rolls Royce in England, General Electric, and Pratt & Whitney. All three are following somewhat similar techniques in attempting to hold noise levels to those specified in the FAA's rules and similar rules imposed in Europe and elsewhere. Rolls Royce, which will provide the power plants for the three-engine L-1011 TriStar, is diminishing noise by several techniques, including reduced turbine fan speed at approach thrust settings, revised nozzle shape, and use of acoustically absorbent linings in the bypass duct and other sections of the engine.

General Electric, builder of the CF6-6 engine which will power the first series of the McDonnell-Douglas DC-10 tri-jet, has developed a glass-fiber sandwich which is being used to line various portions of the engine. The one-inch thick sandwich consists of perforated glass fiber face sheets, a double diamond truss core and solid glass fiber back sheet which acts as a backstop for sound energy. Much research also has been done on installing sound-diffusing rings in the engine inlet ducts, but most engineers oppose using this device because of performance and weight penalties. Every pound of weight added to a jet transport cuts its operating efficiency for airline use.

Pratt & Whitney is developing the JT9D-17 engine, which will provide 48,000 pounds of thrust in the three-jet installation aboard the second series of the DC-10. It will have sound attenuation material installed from inlet tip to the trailing edge of the fan duct, and the inner wall of the fan air duct will be made of standard honeycomb sandwich. This area also may be lined with sound-suppressing material if necessary. Both GE and P&W also are working under FAA cost-sharing contracts totaling more than $1 million for a two-year study of noise-prediction techniques for reducing sound primarily from high-speed air compressor fans.

The new large engines will be noticeably quieter, and other research is underway to find methods of reducing the noise of the existing fleet of jet airliners if and when the FAA regulations are extended to include these. In 1969, under a contract awarded by NASA's Langley Research Center, the Boeing Company conducted flight tests with experimental sound-suppressing engine nacelles on a 707. The engines were surrounded with a high-temperature polyimide plastic honeycomb faced with polyimide laminate, considered most effective for "soaking up" the high-pitched sound from the turbofans during the landing approach. The sound-suppressing devices also included inlets with absorbent honeycomb panels with two concentric sound-absorbent rings in the inlet duct. Longer ducts with absorbent honeycomb walls convey the air aft. The general reaction of 200 test subjects stationed on the ground was that the engine whine of the jets was substantially reduced. The problem, again, is that the sound-absorbent material, though reducing noise, also adds a cost and weight penalty.

The Aerospace Industries Association predicted that compliance with the new FAA regulations would add 14 percent to the direct operating costs of large aircraft. It should be remembered also that devices so far tested are expected only to meet the FAA noise limit of 102 to 108 PNdB. Achieving the FAA's ultimate goal of 80 PNdB around the major airports now seems a virtually impossible task for the industry. However, the outlook is not entirely pessimistic. Rep. Herbert Tenzer of New York, in announcing the Administration's plans to spend $50 million over the next five years to develop a "quiet" engine, said, "I believe we already have the technology to produce a quiet jet engine within a few years. The primary reason we have not reduced the menace of jet noise is the cost factor and the unwillingness of the airlines to place orders for quieter engines. It is time for Congress to put the airlines on notice that we will not stand for further delays."

While the airlines struggled to make economic ends meet, government and industry in 1970 were scheduled to spend some $37.5 million on aircraft noise abatement research. Of that amount, NASA was scheduled to spend $18 million; the Department of Transportation, $5 million; industry about $12 million, and the U.S. Air Force, $2.5 million. Dollar amounts of research do not guarantee results, but at least indicate that public pressure for quiet is being felt in the right places.

.. Other Techniques

The National Aeronautics and Space Administration is concentrating efforts on achieving an engine by 1972 that will give noise-level reduction of 20 PNdB, but in the interim, some relief may come from educating airline pilots and ground controllers to some changes in flying technique. In essence, this effort means steeper descents and ascents on landings and takeoffs and steering flight paths away from most heavily inhabited areas. During 1968 the Boeing Company worked with the NASA Ames Research Center to conduct a program of flight tests using steeper than normal landing approaches. Flying an aircraft with an experimental wing flap for direct lift control, the Boeing test pilots made 150 landings at the Oakland, California airport. They flew a six-degree approach path, compared with three degrees for a normal jet approach, and used lower power settings during the final approaches. The tests were successful, but they pointed up the need for development of advanced direct lift control wing flaps and improved cockpit displays so that pilots may use the technique safely. It is obvious that an airline pilot, if faced with the choice of safety for his passengers or quiet for people on the ground, will choose the well-being of his passengers.

Much of the noise suffered by people living around major

NEW AERIAL GIANTS

airports originates from jet engines on the ground, rather than in the air, either during preflight runups or during tests in the maintenance shop. This particular source of noise can be controlled, at a cost, and some airports are doing something about the problem. Swissair of Switzerland, for example, utilizes individual noise suppressors placed behind the individual engines during ground runups. The suppressors are long-tube mufflers, each weighing 33 tons, and have been shown to be effective in suppressing the ground noise. Los Angeles International is experimenting with a new shield, known as the "hush house" to absorb aircraft engine noise during maintenance checkouts. Forty feet high and 210 feet wide, the hush house is a semicircular structure, open at the top, built of cantilevered fiber glass and steel. It can fit around the rear of any large jet airliner and is effective in absorbing much of the sound which otherwise would radiate to surrounding communities.

The most expensive—and desperate—measure for assisting people who cannot escape noise is to install sound insulation in homes, schools, and business buildings, and even this measure is being tried only out of fear of the rising tide of lawsuits against airport management. The area around San Francisco International Airport is a pioneer in this regard. Thousands of new homes were insulated against sound while they were being built. Offices, hotel rooms, schools, and even the pedestrian malls in neighboring shopping centers were equipped with sound-blocking and -absorbing materials in a campaign started as early as 1960 by an organization known as the Sound Abatement Center.

The insulation of homes was undertaken belatedly by Los Angeles in 1968 when airport officials budgeted $300,000 to soundproof a sampling of twenty-five homes. Techniques include replacing roof shingles with clay and cement tile up to an inch thick; applying fiber mats, gypsum board, and polyurethane foam between joists and in attics; new wallboard hung on vibration-damping spring clips over interior walls; felt-layered doors; double-glazed windows sealed shut; floors strengthened to stop vibration, and sound-absorbing ducts placed in the crawl space under homes. Los Angeles officials also voted to spend up to $225,000 in an experimental program to provide acoustic treatment for schools.

"Busy airports around the world will be watching the results of this research," commented Robert Davidson, general manager of the L.A. Airport. "Even aircraft companies are making great strides in reducing engine noise, but nobody expects them to eliminate noise entirely. I only hope we all can live with it."

Millions of people join him in that hope, but sound-insulating homes and schools proffers another problem. If you open a door or window, the sound muffling effect goes out the window also, and no one can live long sealed inside a closed structure. As Mrs. Georgiana Hardy, a Los Angeles school board member, commented: "A school without windows and with closed doors is educationally bad. And even then a plane could drop on the school." Overriding most other considerations is the cost, and the question of what agency shall pay. The airport and the airlines are being operated both for people and for profit. The homeowner who moves near an airport should share the risk of the noise. Besides the airport operation itself, scores of highly profitable satellite industrial and commercial operations benefit from the air traffic, and these people should bear part of the cost. Paul T. McClure of the Rand Corporation, which assisted in the home insulation experiment around Los Angeles International, said insulating houses *after* they are built costs $2700 to $8900 each. If we were to think in terms of the higher figure —and only one million out of the millions of affected homes— the price tag would total nearly $9 billion. If the law were to place this burden upon airports and airline operators, one of two things would result. Either the airlines would shut down, or effective sound-suppression devices would be found in a hurry. The latter is more likely. Ultimately, such devices will be developed when the airlines refuse to buy unsatisfactory equipment from the aircraft manufacturers.

.. Land Use Planning

One partial solution to airport noise is proper land use, but unfortunately this essential planning function is probably the greatest weakness of American municipalities today. The theory is simple: remove the people from the airport, or the airport from the people. It should be said in favor of fumbling city government that most airports originally were located well beyond city boundaries, far enough away that few people were disturbed by the noise. This was done, however, not with noise planning in mind but for flying safety, and once the airports were established, the communities did not follow through with zoning laws which would provide adequate buffer. Most cities, as the airports themselves generated commercial business, allowed office buildings, homes, restaurants, schools, and housing developments to swarm in around the airports in willy-nilly fashion. Soon, as in the case of Midway at Chicago and Los Angeles International, thousands of homes and millions of people have crowded in virtually to the edge of the runways.

It is acknowledged that intelligent zoning is one of the most difficult problems encountered by city officials, because zoning frequently involves conflict with an individual's right to use his property in the most profitable manner. However, the only practical way to prevent people from being annoyed by aircraft noise is to ascertain that they will not live near the airport. Given a situation in which a new airport is being planned, acoustic engineers can draw realistic noise profiles on the surrounding countryside based upon the number and direction of runways and the anticipated air traffic load. In general, it would be best to prevent homes from being built under the approach and takeoff paths of the main runways for a distance of at least three miles from the end of the runway. Also, no homes should be permitted in an area at least one-half mile from each side of the airport or its runways. It can be seen that a considerable amount of land is involved, perhaps two or three times as much as the land required for the airport itself. Ideally, a city should buy this additional land so it could maintain direct control, but most municipalities are unwilling, too shortsighted, or unable to do so. The alternative is to wait until the noise problem becomes unbearable over built-up properties, which then may require purchase by the city at much higher prices. With proper zoning, the land within the noise contour profile may be put to excellent use for hotel, commercial office, or industrial purposes—any highly profitable use wherein an economic justification may be made for adequate soundproofing *when the new buildings are being constructed.* No matter what approach is taken, someone must pay the bill—the city, the state, the airlines, the home-owner, or a combination of all. Unfortunately, sound land use planning can only be useful now in the planning of new airports. The old ones already are surrounded by people whose rights must be protected regardless of the lack of wisdom which permitted them to live in the direct noise path.

Some cities, because of the inexorable growth of air traffic, today are being forced to plan new airport facilities or expand old ones and thus are in a fairly good position to put regional land use planning into effect before it is too late. Some of these are Montreal, Canada, Cleveland, Ohio, and Los Angeles.

At Montreal, Canada's Department of Transport required that a wide belt of nonresidential property be set aside around the new airport site at Sainte-Scholastique. Cleveland's proposed new $1.2 billion jetport, suitable for the jumbo jets and supersonics of the future, is designed for location a mile offshore in Lake Erie. Scheduled for completion by 1978, the jetport is designed to serve a region encompassing 22 million people until the beginning of the next century. In a bold move to pull the big jets of the future away from the congested metropolitan area, Los Angeles has reached across a range of mountains to a desert location near the community of Palmdale east of the city. The big intercontinental airport there is to be master-planned from the beginning so that there will be no residential encroachment. This, plus a growing network of interrelated airports, is expected to serve 250 million travelers per year by 1985. William L. Pereira, architect planner for the $700 million L.A. expansion program, predicts the Palmdale area will become the first major metropolis of the air age. The airport area, he said, will be "sterilized to exclude uses incompatible with supersonic transport noise. Areas will be zoned for recreation, industry and open space—not homes. The problem of aircraft engine noise is being studiously pursued and eventually will be licked, but there is no question, this metropolis will be different."

... The Threat of VTOL

While the nation comes to slow grips with the airport noise problem, an even more serious threat of aircraft noise looms in the immediate future. This threat is the Vertical (or Short) Take-off and Landing Aircraft (abbreviated as VTOL and STOL). Although the helicopter has been in limited service, primarily with the military, for many years, an entire new generation of STOL and VTOL aircraft is nearing the operational stage. Their arrival on the aviation scene is at an appropriate time because major metropolitan centers have allowed their surface bus systems to decay into uselessness under the choking onslaught of the automobile. Although the situation has been deteriorating for many years, the cities now find themselves generally unable to cope with the task of moving large numbers of people short distances—such as between two neighboring cities, or from a busy airport to the downtown city center.

The VTOL and STOL offer part of the answer to this problem. Rising and landing either vertically, or with a runway distance of only a few hundred feet, these aircraft will, it is anticipated, operate from city center to city center. In other words, many cities soon will have downtown landing pads, ranging from building tops for helicopters to four-block-square fields for other types of planes. These aircraft will not require extensive clear space around the field for takeoff or landing approach. Some will be helicopters, but a number of other types also are in advanced design and test. One type has propellers installed in the wings for takeoff and landing. Others have wings that rotate, with propeller-driven engines, for both vertical and horizontal flight. And at least one jet plane now operates in this fashion. This is the British Harrier, which was most dramatically demonstrated during the 1969 transatlantic race from the heart of London to the top of the Empire State Building in New York City. The Harrier, after crossing the Atlantic, landed closest of all the participating aircraft to the finish point of the race. It turned its jets downward and landed on the STOL-port which has been built at the edge of the Hudson River near the towers of lower Manhattan. The remainder of the trip was by cab or subway.

There is no question that these aircraft will serve a vital public function, particularly as our metropolitan centers grow to megalopolis size. In the absence of truly rapid ground (or subterranean) transit offering travel at 100 miles an hour or more, the VTOL may become the only alternative to the automobile in moving 50 to 200 miles. Some cities, including Los Angeles, have contemplated a massive helicopter-bus combination which could transport entire busloads of people from an airport to a downtown center by air, and then travel by surface street to more specific destinations.

The concern is very real that with VTOL and STOL, the airport community noise problem will be carried directly to the center of our major cities, ranging from the clattering sound of helicopter to turboprop to pure jet.

The aircraft industry is aware of the difficulty. In a paper presented before the American Institute of Aeronautics and Astronautics, two Lockheed Aircraft experts, Nathan Shapira and Gerald J. Healy, warned that "a new form of aircraft noise may be expected as the ever-worsening ground traffic congestion is relieved by short-haul air transports. The anticipated growth of the short-haul air traffic will bring vertiports into established commercial and industrial areas, and with them, the noise of low-flying aircraft will be intensified over residential communities. The potential problem of short-haul air traffic noise must be assessed realistically now when foresight may prevent a problem." The two engineers urged "judicious site selection [for vertiports] and the effective use of zoning and building code regulations" as a partial prevention of noise problems.

Thus a straw of hope is in the wind that the noise problem in this case is receiving at least some consideration before it reaches unbearable proportions. Downtown VTOL pads could be flanked by high structures, such as office buildings. On the side facing inward, of course, it would be necessary to completely soundproof (and accident-proof) these buildings. On the noise source end, some experimentation also is being considered to quiet the STOL engines. A Canadian (deHavilland DHC-7) turboprop is utilizing propellers of 11.5 foot diameter for lower tip speed (and thus lower noise). Sweden has established stringent aircraft noise limits near cities, and the Saab 1071 turboprop was designed with low noise as a prime consideration. In the United States, Garrett AiResearch is working with NASA and the FAA to develop sound-suppression techniques for the turboprop engine. Devices under test include low-speed propellers, variable-speed airbox, bleeding air through the propeller to break up noise generated by eddy currents at the blade's trailing edge, and sound absorbers in the exhaust system.

Again, as with any war of attrition, victory over aircraft noise will be the aggregation of hundreds of tiny advances, each subtracting a fraction of a decibel from our fractured ears until hopefully, someday, it may be comfortable to live near an airport.

While that is happening, however, the supersonic aircraft threatens to carry the noise problem beyond community limits, spreading in wide belts across the continents and the oceans of the world.

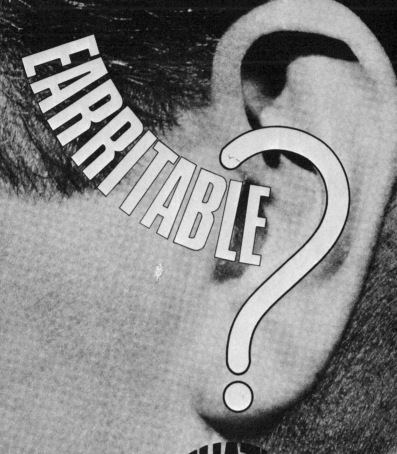

EARRITABLE?

THAT'S HOW NOISE GETS YOU

help keep noise DOWN

Government Information Office in co-operation with The Noise Abatement Society   Printed in England by Eyre and Spottiswoode Limited, at the Grosvenor Press, Portsmouth

CHAPTER 8

*Would the $3 billion worth of time
saved for busy executives by SST
flights over North America
repay society for the more than
$80 million in damage claims
such flights would probably cause?*

The Sonic Boom . . .

FROM JANUARY THROUGH March 1965 the City of Chicago was subjected to a thunderous bombardment (minus bombs of course) by the U.S. Air Force. The rolling booms of thunder which swept across the city forty-nine times during the three-month period were created by the USAF B-58 four-jet bomber, first supersonic bomber built for Air Force service. The B-58, additionally, was designed specifically for low-level radar bombing runs at supersonic speed, a tactic designed to confound enemy radar if the United States were ever called upon to bomb an enemy nation.

The B-58 is not an especially large aircraft, although much larger than the Air Force fighter aircraft which, prior to 1961, were the only planes making supersonic runs over any part of the United States. The size of the bombers—and to some degree their shape, but most especially their low level of flight—created sonic booms of heavy overpressure. The Air Force, as a result, was flooded with 3156 damage claims filed by citizens who had suffered broken windows, cracked plaster, and structural damage to their homes and business establishments, as well as some who claimed personal injury to themselves and animals.

The 3156 damage claims resulting from the B-58 sonic booms of 1965 are only a fraction of the more than 35,000 sonic boom damage claims which the Air Force processed between 1962 and 1968. That number, in turn, is only a small fraction of the millions of people who have been startled and annoyed when the windows of their houses vibrated like a drumhead before the explosive waves of air pressure which sweep across the land in the wake of an aircraft flying faster than the speed of sound.

If any noise in our environment exceeds that of the sonic boom, it may well be that generated by people who oppose the boom. Therein lies the greatest hope that the latest child of technology, the supersonic transport plane, will not be permitted to disturb its neighbors. The neighborhood, in the case of the SST, reaches beyond the borders of any jetport to include every person who lives beneath the well-traveled airways of the world.

There is a great deal of both fact and fiction about the sonic boom. On one side, statistical experience has proven that the boom can break window glass, damage plaster walls, and cause structural damage to very old buildings such as historic structures and monuments. On the other hand, a great deal of damage claimed to have been caused by sonic booms was caused by other forces. There is no record of human injury by the noise or its pressure wave, but the sonic boom is undeniably a shock to the human nervous system, and that shock reaction alone is reason enough for prohibiting the supersonic aircraft which causes the boom. Air Force claims files contain cases ranging from a frightened dog killed by a passing car to chicken eggs that did not hatch.

Some of these claims are valid; most are not. During 1968, the Air Force conducted careful investigation of a series of claims that sonic booms frightened mother mink, on domestic mink farms, causing a great loss of production during whelping season in April and May each year. In June 1968, Air Force investigators were confronted with twelve mink loss claims totaling nearly $310,000, and although a series of tests simulating the boom failed to prove the effect on mink, one case was settled for $31,000 and additional studies were undertaken. "Present policy," commented a high-ranking Air Force officer, "is to avoid overflight of mink ranches during whelping season."

.. The Boom: Pro and Con

In 1968, then Interior Secretary Stewart L. Udall established a national committee of select scholars to conduct a thorough study of the sonic boom in terms of its potential dangers to man. This group, while sharing the general concern over the environmental degradation threatened by noise of this caliber, also attempted to come to grips with the extreme positive and negative points of view which have grown up around the question.

"The subject of the sonic boom," the committee reported to Udall, "has generated a great deal of controversy, much of which seems to be based upon hearsay, emotion or prejudice rather than upon fact, logic or analysis. The spectrum of opinion spans a wide range. One end is rooted in the view that the creation of new technological hardware is itself a primary goal, and that man should and will adapt to conditions imposed by the technology and its use. This view recognizes that there may be undesirable effects, but assumes that man's adaptability and reasonable regulations of use will successfully cope with these effects.

"The other end of the opinion spectrum has its base in the concept that man already is deeply insulting his own environment. This view holds that man's fouling of his environment has within it the seeds of his downfall; that it is essential to rehabilitate the environment, and that no new insults should be permitted. This view generally maintains that certain new technological advances carry threats which entitle them to be placed in the category of poisons, for which the only appropriate regulative posture is their use only for defense purposes.

"Somewhere in the middle ground," the committee said, "is the opinion that the use of new technological creations should be subject to enlightened judgment, balancing the economic and social advantages against the social and environmental disadvantages. This process of judgment requires the free presentation of facts, open analysis, public discussion of ideas and opinions, and the search for salient data to undergird the decision-making process. In the case of the supersonic aircraft and its use, with the sonic boom inherent in its operation, there appears to have been insufficient availability of facts on which to base analysis and balanced judgments."

Among the adamant, and sometimes eloquent, voices sounding out against the sonic boom is that of Dr. Garrett Hardin of the University of California at Santa Barbara. The boom, he states, "is something much worse than noise. Experiencing it is like living inside a drum beaten by an idiot at insane intervals. Disconcerting is the mildest word we can use for this experience. For invalids, for the mentally disturbed, for surgeons surprised in delicate operations, for those who want to sleep or pray . . . such a disturbance is scarcely tolerable. Every supersonic transport flight from coast to coast will disturb 20 million groundlings as it enables 400 passengers to save three hours apiece. Is it worth it?"

Another is W. H. Ferry, previously cited, who has headed the Citizens League Against the Sonic Boom. "I believe," Ferry stated, "that the case against noise must be made on the basis of civilized standards as much as on the basis of public health or economy. If research, for example, should definitely establish that no human or material harm was being done by the sonic boom, I would recommend a gigantic increase in the campaign against it, because it is de-civilizing. No self-respecting civilization ought to have to accommodate itself to such an annoyance. We should not have to produce irrefutable evidence that our health is being impaired before action can be taken against the more and more numerous assailants of quiet. Economic benefits should not be conclusive in any argument between noisemakers and people. Civilized life means amenities, among which quiet and privacy rank high; and I look on civilized life as one of the principal goals of all activity here below."

The sonic boom itself is not a new phenomenon. The sound is well known to any soldier who has been subjected to artillery or naval gunfire. More than a half-century ago it was axiomatic among the men in the trenches of France during World War I that the shell which could be heard booming overhead was not the one that would hurt you. It was not until a generation later, in the waning days of World War II, that the fastest of propeller-driven planes, during diving situations, approached speed which later became known as the sonic barrier. This speed is the speed of sound itself, which in the atmosphere varies from 760 miles an hour at sea level (60° F) to 660 miles an hour at 65,000 feet and −70° F. Current subsonic jets cruise at 30,000 to 40,000 feet altitude. In 1949 the U.S. Air Force experimental plane, Bell X-1, achieved a speed of 940 mph, and by 1959 the military forces of the world were operating jet-powered fighter aircraft which could exceed the speed of sound in normal level flight. The sonic boom, as we know it, thus was born.

An airplane, in subsonic flight, radiates a sphere of sound which dissipates in all directions. In supersonic flight, however, the sound waves are not able to escape in front. The air compresses, pushing a shock wave before it. This three-dimensional shock wave spreads out in a cone shape behind the aircraft (with the plane at the point of the cone) similar to the two-dimensional wave of water breaking back from the bow of a ship. Where the bottom trailing edge of this shock wave strikes the ground, it produces the sonic boom. Although the boom is heard as a momentary burst or rumble of sound at any one point, the wave travels along the ground beneath and behind the plane. The sonic boom often is produced by planes flying so high that they are invisible to the naked eye and the sound of their engines is inaudible. A low-flying supersonic plane will produce a narrow pressure band of high intensity. A plane flying at 60,000 to 70,000 feet will cause a much milder boom due to distance and atmospheric attenuation, but the trace of the shock wave on the ground would cause a swath of moving sound 60 miles wide. In other words, a supersonic plane traveling across the United States would produce a sonic boom audible to every person living in an area 60 miles wide and 3000 miles long. This is why the advent of the supersonic transport has been so vigorously opposed.

The boom, when it occurs, usually is two in one—two explosive bursts of sound in quick succession. This is due to the peculiar N-shaped signature of this particular type of pressure wave. The sound first moves abruptly to a peak above normal

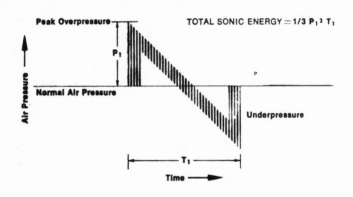

THE SONIC BOOM. Idealized boom recordings resemble an N and show the rise and fall in atmospheric pressure. The energy of a boom depends on both peak pressure and duration.

—*Reprinted from* Psychology Today *magazine, November 1968,* © *Communications/Research/ Machines/Inc.*

atmospheric pressure (the top point of the "N"), and the second burst occurs at the lower point of the "N" when the pressure drops below atmospheric normal. The force of the boom (which may equal sound intensity up to 130 dB) is generally measured in pounds of pressure per square foot. In the case of high-flying aircraft, this normally would vary from one to five pounds pressure per square foot. Comparing this with normal atmospheric pressure of about 14.7 pounds per square *inch,* it is evident that the sound pressure of the sonic boom (discounting rare and exceptional cases) is not sufficient to harm the human body or hearing mechanism. Part of the reason for this, despite the high decibel equivalent, is that most of the sonic boom occurs in the very low sound frequency range. The sound may vary from a sharp, explosive noise which rattles walls and windows to the sound of low thunder in the distance. The time span of the sonic boom ranges from about one-tenth to one-half second, depending upon the design, size, speed, and altitude of the aircraft as well as wind direction and other atmospheric conditions. A person standing below a supersonic transport plane traveling at 1800 miles an hour would hear the sonic boom as a sharp crack, a thump, or distant rumble.

Researchers have found extreme difficulty in measuring the intensity of sonic booms, because of the variability of the sound. Beyond that, it has become virtually impossible to establish standards of how much sonic boom will be tolerated by the public without serious complaint. Residents around Air Force bases, where the sonic boom has been accepted as part of the environment over a long period of time, tend to be less annoyed than metropolitan dwellers who are subjected to the boom. The difference is partially due to the *meaning* of the sound. In the first instance, the boom has come to mean that people are being defended by fast, up-to-date aviation equipment; and the noise therefore is tolerable. On the other hand, people would not feel the boom justified if it meant nothing more than a faster method of transportation. However, the question of how much sonic boom is enough is being forcefully answered by the people who want none of it at all.

The Air Force, which has conducted supersonic flights on a routine operational basis since the beginning of the 1960s, has accumulated the largest body of statistical knowledge regarding public reaction to the sonic boom. As a result of that knowledge, Air Force pilots observe standing orders not to exceed the speed of sound over populated areas unless flying at 30,000 feet altitude or above, and to avoid flying over heavily populated areas whenever possible. Since Air Force attorneys have investigated thousands of sonic boom damage claims during the past ten years, they form the greatest accumulative authority on what damage may actually be caused by the sonic boom. Even here the evidence is elusive because, as one Air Force report states, "sonic boom damage is often 'discovered' rather than seen to have happened." Many claims are filed covering damage completely unrelated to the sonic boom.

"Though occasionally an eyewitness may see a window break or observe a plaster ceiling fall," the report states, "most sonic boom damage is found either (1) after the claimant hears the boom and begins inspecting his property thoroughly for damage (often for the first time in years) or, (2) when the claimant suddenly perceives damage which may have existed for a considerable time and then searches his memory for a recent event which he feels could have caused the damage. In the latter instance, the damage could well have existed for years and sometimes claims investigators, upon close inspection, find dirt or even paint in the newly found glass or plaster cracks. However, the unperceptive property owner often will remain unconvinced and will continue indignantly to blame the sonic boom which had recently startled or annoyed him."

"Indeed," the report stated, "it is suspected that there is a very important factor in the subjective attitude of the property owner toward government, aircraft or other items disturbing his tranquil existence which has an effect upon his propensity to make claims for progressive property damage which might be attributed by other less affected property owners to natural effects." This "subjective attitude" of the property owner is really the point of the entire argument. Air Force investigators logically resent false claims, but they should remember that no claims at all would be filed unless a sonic boom had occurred somewhere.

Although thousands of such claims are false, the Air Force tends to give the benefit of the doubt to the claimant in cases where the sonic boom was, or could be, at fault. Most common such cases are broken windows and the collapse of wet plaster before it has had time to set. In the past there has been an occasional accident in which a low-flying aircraft broke the sound barrier at low altitude, causing considerable damage under its path. One of the most embarrassing such incidents occurred in recent years at the Air Force Academy near Colorado Springs, Colorado, when a pilot, seeking to impress high-ranking officers during a ceremony, accidentally buzzed the campus at supersonic speed. The boom shattered large plate glass windows in one of the beautiful Academy buildings and several people were injured by flying glass.

Most complaints and damage claims, however, originate in areas where there is no clear-cut record that a sonic boom has occurred. Researchers therefore welcome specific tests from which results may be correlated with accuracy, both as to damage and annoyance to people beneath the sound pressure wave. In addition to the Chicago B-58 overflights mentioned at the beginning of this chapter, another such case occurred in 1967 when the SR-71, America's fastest military aircraft, began high-level reconnaissance flights on regular routes throughout the United States to test its photographic sensor equipment. Again Chicago citizens provided a sampling of test victims. The SR-71 flew over the city at supersonic speeds and high altitude 51 times from July 30 to October 2. The claims office at nearby Chanute Air Force Base, Illinois, received 2309 telephone and written complaints of damage. Of these, 391 followed up by filing correct claim forms. Finally, the Air Force approved 95 of the claims (with payments totaling $6470), 239 were disapproved, and 11 were withdrawn. Overall, the Air Force has found that proven sonic boom damage includes 65 percent glass breakage, 21 percent plaster damage, 8 percent damage resulting from falling objects, and 6 percent miscellaneous. Single-family structures represented about 70 percent of the claims, with only 10 percent occurring in multifamily units. An interesting fact emerging from Air Force statistics showed that in buildings as high as 40 stories, almost all of the claimed damage occurred on the ground floor.

While legal determinations seem always to be limited to physical damage, there is no record of the millions of people who suffered severe annoyance yet considered themselves helpless to recover damages or stop the supersonic flights over their heads and their homes. Although physical damage resulting from sonic booms may have been exaggerated, nervous tension may be aggravated by the unpredictability of the sound and the frequency with which it occurs. As Mr. Ferry has made clear on an earlier page, the case against the sonic boom (and any noise) must be made on the basis of civilized standards as well as upon public health or property damage. There is probably no statistical

THE SONIC BOOM

measurement that could be made of a man's mental health deterioration if he fears his home is being gradually shaken to pieces by the sonic boom, whether or not this is really happening.

The United States, and other governments, have conducted a number of research studies subjecting population centers to measured numbers of actual sonic booms. One six-month series was conducted over Oklahoma City in 1964 and another in 1966 over communities around Edwards Air Force Base, California. The latter studies used 300 test subjects to provide subjective ratings comparing the sonic boom with the loudness of a subsonic plane flying overhead. For indoor listening, sonic booms of 1.69 psf (pounds per square foot) overpressure were rated at a level equal to a 109 PNdB subsonic jet noise. In general, the sonic boom was more acceptable to outdoor listeners. The difference is attributed to the fact that when a person is indoors, he is alarmed by the rattle and vibration in his building even though the boom may not appear to be as loud. The various studies have shown that a substantial portion of all people would find it unacceptable to be bombarded eight to ten times a day with sonic booms of 1.2 to 1.7 psf overpressure. In the Oklahoma City study, 27 percent objected; in St. Louis, 35 percent; at Edwards Air Force Base, 26 percent, and in a series of studies conducted over French towns and countryside, 34 percent of the people complained bitterly.

While statistical curves are being refined to determine the maximum sound the public will bear, the public is making enough noise to force politicians, industrialists, and airline operators to pause for a moment in their headlong rush to bring the next technological revolution in the air.

... The Supersonic Transport

In his inaugural address in January 1969, President Richard Nixon stated: "We find ourselves rich in goods, but ragged in spirit; reaching with magnificent precision for the moon, but falling into raucous discord on earth. . . ." A good share of that raucous discord has to do with the growing awareness among 205 million Americans that our environment has seriously decayed before the thoughtless onslaught of air, water, land, and now noise, pollution. More raucous discord is yet to come on the question of whether America will build and fly the supersonic transport.

This aircraft, which will more than double the speed at which men may move from one point to another on the surface of our planet, is the next logical step in the development of air travel. That development already is well underway, justified by industrial and political forces on the premise that the SST is needed to maintain our economic and competitive position among the industrial nations of the world, several of which already are well advanced in supersonic transport technology. Opponents of the SST challenge its need, when the economic and competitive advantages are weighed against the additional environmental decay it would cause. Many are asking, again, if the blind rush of technology should not slow in its mad pace while the factor of human amenities is cranked into the equation of engineering design. In the words of famed physicist Max Born, who died in 1970 at age eighty-seven, "Intellect distinguishes between the possible and the impossible; reason distinguishes between the sensible and the senseless. Even the possible can be senseless."

There is little doubt that the supersonic transport is possible. Time will tell if it is senseless. While we wait for the answer, the British and French are flying the Concorde and the Russians are testing the Tupolev 144, both designed for transport of passengers at supersonic speeds. The Boeing 2707, tied to a cooperative venture with the U.S. government for development funds, cannot fly commercially before the 1976-78 time period. During that time, many questions will be answered, including those of sonic boom and engine noise control.

The Anglo-French joint venture on the Concorde began in 1961 and resulted in a 375,000-pound aircraft which can carry about 120 passengers at 1200 miles an hour. It will barely be able to accomplish nonstop transatlantic crossings fully loaded. The Russian TU-144 appears to be similar to the Concorde, though perhaps 100 miles an hour faster.

The American SST project has been considerably more ambitious in size, speed, and range since its inception. It began, officially, in 1963 when the FAA issued a request for proposals from

U.S. airplane engine and airframe manufacturers. By early 1964, most firms had been eliminated and definitive competition narrowed down to the Boeing Company and Lockheed for the airframe and General Electric and Pratt & Whitney for engines. In 1966, Boeing (with a controversial swing-wing design) won the airframe competition and GE was chosen to develop the engine. The FAA put up 75 percent of early development funds, which were expected to exceed $1.5 billion by the time two prototype planes could be built.

Since that date, Boeing has made drastic design changes. Because of money, noise, and political problems, among others, the go-ahead for prototype construction was delayed into late 1969. The Boeing 2707, when it flies, will be capable of carrying 298 passengers at 60,000 feet altitude and 1800 miles an hour. Although many people still wonder why anyone would want to fly that fast, here is the kind of service the 2707 could provide: A person could leave New York City at 6:00 A.M., arrive in Paris at 3:00 P.M., spend the afternoon and evening there, return on a flight leaving Paris at midnight, and be back in New York by 9:00 P.M. the same day he started. Total elapsed time for the trip would be 15 hours, with 9 in Paris. In the other direction, a round trip leaving Los Angeles at 8:00 A.M. and returning at 11:00 P.M. could provide a full business day in Honolulu. With the SST it will be possible to fly halfway around the world in any direction in half a day. But again, with the delays which democratic debate have built into the American program, it will be five to eight years before the Boeing SST can enter commercial service. Supersonic speeds would greatly enhance U.S. commercial ties with the nations of the far Pacific, but the question of what such a combination of speed and distance traveled will do to the physical and mental balance of travelers is yet to be answered.

After a decade of studies concerning economic feasibility of the SST, Boeing marketing experts predict that before 1990, the SST will be carrying more traffic (at least in international service) than all the current Free World air travel. Revenue passenger miles are expected to increase from 430 billion in 1975 to more than a trillion in 1985. The SST is expected to relieve airport and airway congestion by spreading traffic over a greater portion of the day, and an SST will occupy air space only one-third as long as a subsonic jet.

Projections of current design indicate that the Boeing 2707 will produce a maximum sonic boom of about 3.5 psf over-pressure during climb, dropping below 2 psf during cruise and increasing to 3.5 again during descent to subsonic speed. Aside from the boom, the SST's more powerful engines may produce noise around airports up to 124 PNdB at distances up to 1500 feet on each side of the runway. To bring this in line with proposed airport noise limits, Boeing anticipates it may be necessary to add up to 18,000 pounds of sound-suppressing material on each aircraft. This question of engine weight, efficiency, and noise control may well be the deciding factor delaying the American SST even farther into the future.

The high-level committee which studied the SST problem for the Interior Department found that if commercial SSTs are allowed to fly at supersonic speeds over the continental United States, the expected frequency and intensity of sonic booms would represent a significantly large increase in the noise level, and in the numbers of people exposed to intense noise. "The numbers of SSTs expected sometime after 1975," the committee reported, "would subject between 20 million and 40 million Americans under a path 12½ miles on either side of the flight tracks to 5 to 50 sonic booms per day. Each boom would be perceived by its hearers as equivalent in annoyance to the noise from a large truck traveling at 60 miles per hour at a distance of about 30 feet, or a four-engine turbofan jet airplane within half a mile of its takeoff point. An additional 35 to 65 million people within 12½ to 25 miles of the flight path would be subjected to one to 50 booms per day of somewhat lower intensity, and 13 to 25 million more would experience 1 to 4 high intensity booms. The response of the people in the 25-mile-wide swaths swept by frequent and intense booms can be expected to be similar to that of residents of neighborhoods adjacent to busy metropolitan airports under the flight paths of planes taking off."

Members of the committee pointed out that complaints and damage claims derived from sonic booms already have forced restrictions on the flights of military supersonic aircraft over

THE SONIC BOOM

populated areas of the country, although the boom from military jets is less severe than that anticipated from the larger transports. The report estimates that regular commercial overland flight across the United States would produce from 3 to 6 million damage complaints per year, with one out of three or four of these resulting in award of damages. A conservative estimate of the continuing annual cost of the repair of damages to houses and other structures (not counting the cost of processing claims or inspection of damages) is at least $35 million and probably more than $80 million per year. Against this cost (and there is no clear-cut indication of the agency which would be responsible for paying it) the committee estimated the value of time the SST would save for "busy, highly-paid persons who would probably be the majority of SST passengers" might amount to $3 billion per year. The moral question remains whether the people who benefit from the SST could have the right, at any cost, to inflict damage and annoyance upon the other millions of people in the country. That question must be balanced against a worldwide market for the supersonic transport which may be as large as $20 to $30 billion between now and 1990.

Aside from the millions of lawsuits the sonic boom would produce, what of the peace of mind of the victims? As expressed by Stanford Law Professor W. F. Baxter: "All the cracked $5 windowpanes, all the dinner dishes dropped on kitchen floors as the result of startled reactions, all the millions of hours of sleep lost while frightened children are comforted, the razor-nicked chins, the interrupted concerts, the hammered thumbs, the fallen cakes and omelets—all these will produce not litigation but a silent curse upon the industry, the Federal Aviation Administration and a society that seems to have confused technology with civilization."

. . . Boom Reduction

What, if anything, can be done to reduce the shock of the sonic boom? A number of techniques may make it milder, and future research may eliminate the problem entirely, but for the moment there is no cure except prevention.

In aircraft design, a swept-back wing produces a milder shock wave, since boom intensity is dictated to a great extent by the airplane cross section. In general, a large aircraft produces a greater shock than a small one. Among futuristic ideas for reducing the boom, Edwin L. Resler Jr., director of the Cornell University graduate school of aerospace engineering, believes the shock could be cut in half without radical change in aircraft design. He suggests redesign and location of engines so that the stream of air leaving the engine would be smaller than that going in. This would create a partial vacuum behind the aircraft, sucking in some of the cone of shock waves and dulling the impact far below on the ground. Engineers at NASA's Langley Research Center theorize that if an SST were lengthened by 50 percent, the boom would be reduced by one-half. Professor Raymond A. Bauer of Harvard University believes the boom may be reduced by flying at higher altitude and by incorporating slight design changes. Two researchers at Northrop Corporation, where aircraft noise reduction has been under study for many years, proposed the theoretical possibility that the boom could be diffused by generating an electro-aerodynamic field ahead of the SST in flight. As proposed, a corona discharge over the nose and wing of the plane would deflect and weaken the leading edge shock wave which produces the boom. Although the workability of this thought has been disputed, partially because of the high weight of the required onboard generating equipment, it is probably an offbeat idea such as this which may lead to the ultimate solution of the sonic boom problem.

While technology labors to reduce its own noise, many nations of the world already have passed regulations prohibiting SST overflights at supersonic speed. Continued pressure by the U.S. public is needed to assure that the same restrictions are imposed here until the SST is tamed to the point where it will not disturb its neighbors more than it's worth. So far, public pressure has had the desired effect. As late as January 1970, Secor Browne commented: "The supersonic transport, in my opinion, because of the boom will be restricted in operations. It will not fly at supersonic speeds over articulate voting populations."

THE SONIC BOOM

The Boeing Company by late 1969 had shifted all of its SST market studies to the assumption that the SST will fly supersonically only during transocean crossings, or north of the Arctic Circle where few people would be harmed or alarmed. Market analyses show that 80 percent of the world's intercontinental air seat miles are produced over water, and Boeing's studies showed that even if all continental flights are conducted at subsonic speed, the SST continues to promise high market potential. Under these restrictions, Boeing foresees demand for 515 SSTs by 1990.

Emerging from all of the sonic boom debate around the world is the persistent hope that at last technology is hearing the public above the noise of its own machines and may indeed aim toward a quieter world.

THE CASE
FOR TRANQUILITY

CHAPTER 9

*The insidious nature of noise-induced
hearing loss has made it difficult
to track down industrial culprits
and to justify including such injury
under workmen's compensation laws.*

The
Industrial
Din . . .

THE INDUSTRIAL REVOLUTION, powered originally by the coal-fired steam engine, introduced the human race to the age of man-made noise. It introduced us also to immense material progress while starting us upon the road to degradation of our entire environment—air, water, and land. Two centuries later, power sources and machines are more sophisticated, but the greatest danger to human ears still remains the noise generated within the walls of major industrial plants. The primary reason is that machines are invented and built to do a specific job. So far in our history, the economic function has overridden the consideration of physical and mental injury which machine noise causes.

This is not to say that ear- and nerve-shattering noise has been ignored in industry. In fact, the opposite is true. Industry has provided us with the greatest statistical laboratory and controlled environment for studying the effects of noise upon men and women. It is in industry that noise has been measured in terms of profit and loss, in terms of work efficiency, and finally in terms of injury to individual human beings.

Literally millions of hearing measurements in industrial situations have been made during the past half-century. Correlations by Dr. Aram Glorig, one of the nation's foremost authorities and tireless investigators in hearing loss and conservation, show that people not exposed to industrial noise have hearing which, throughout the age spectrum, averages 10 to 30 percent better than that of industrial workers. One broad study of 55-year-old people shows that among those who have lived in quiet places all of their lives, only 22 percent have suffered significant hearing impairment. However, in the same age group of those people who have worked near noisy industrial machines for many years, 46 percent, or nearly half, have suffered serious hearing loss.

Thus, since 1830 when hearing specialists first were able to document noise as the cause of hearing loss among blacksmiths in England, we have compiled massive and conclusive evidence that loud, sustained noise, particularly in an enclosed space, does serious damage to the human hearing mechanism. Although evidence of mental or psychological damage is harder to document, industry also has been the proving ground for the allegation that noise causes a serious inefficiency among workmen. Although that fact alone has been the motivating force behind many industrial efforts to reduce, alleviate, and eliminate noise sources in the work environment, another force has hastened the process. This is the legal principle which requires that a human being must be compensated for work-caused illness or injury.

It has long been recognized that if a man lost a hand in a metal cutting press, or was blinded by a flying scrap of steel from a lathe, he must be compensated for the loss. However, application of this principle was slow in coming to industry where hearing loss is involved because of the intense difficulty of proving the *source* of the hearing loss. However, during the past 50 years, the courts and legislative bodies have gradually come to recognize industrial hearing loss as compensable. A workman whose hearing has been seriously impaired because of intense noise on his job must be paid for that loss now under the law in most of the United States.

It is this foundation of law, born in the courts and ripening for the past half-century, which forms our most substantial hope in the future for defense against the uncontrolled proliferation of noise in our general environment— in cities, on the highway, around our airports, and in our homes. Once it is clearly established that the person or agency causing the noise—be it a truck, bulldozer, supersonic jet, or jackhammer—must *pay* for the personal injury created by that noise, then and only then will our world become quieter. It is a common failing of human nature for a man to seek personal profit from his labor and machines without taking the needs of other people into consideration. However, when the cost of noise in hard dollars begins eating into the profit of a manufacturer or an airline, then noise abatement will be taken under consideration when the machines are being designed, not after they already are in operation. As illustrated previously, the case of the supersonic transport probably is the first time in our history that a public protest may have throttled a source of aggravating and injurious noise *before* it became a *fait accompli*.

Overall, industry in which men have harnessed machines to do their work is the fundamental source of most of our noise today. It is likewise the foundation point from which tomorrow's law may protect us from a universal loss of hearing and perpetual noise-induced nervous tension. Because industry is the source of most man-made noise, it is necessary to examine briefly the long struggle which preceded the legal recognition that a man's hearing must be protected on the job, or else he be paid for its loss. Progress has been slow in this area, but it is coming.

Major industries in which hearing hazards exist include iron and steel making, motor vehicle production, aircraft manufacture, textile factories, paper making, metal products fabrication, printing and publishing, heavy construction, quarrying, mining, lumbering and wood products, and mechanized farming. The actual list is far longer and might include, for example, Armed Service occupations such as flight line and carrier deck operations, engine test cell and weapons firing, armor operations, and assorted repair and maintenance work.

The Committee on Environmental Quality of the Federal Council for Science and Technology estimated in 1968 that the number of United States workers now experiencing noise conditions unsafe to hearing is more than six million. It may be as high as 16 million.

"With some exceptions," the Committee wrote, "verification of hearing loss in different occupations has been difficult owing to management's fears that such tests might precipitate an avalanche of compensation claims. Unions have not pressured for such surveys either." The only logical way to assess hearing loss due to industrial noise is to give a hearing test to each person when he first reports for work in a particular factory. Later, if he should file a claim for hearing loss, the evidence at least would be clear concerning the amount of loss he had suffered, whether or not it was caused by his work environment.

The potential cost of compensation for industrial hearing loss is alarmingly large. The twelve-man Committee on Environmental Quality estimated a potential cost of $450 million per year, assuming that only 10 percent of those eligible for hearing loss compensation would file a claim and that the average award per claim would be $1000. In actuality, hearing loss awards average $2000. While more and more hearing loss claims are being processed each year, the total number is still relatively small. Indications are that fewer than 500 cases were settled in 1966.

There are various reasons for the small number of claims. Many afflicted workers do not know they can claim payment for their hearing loss. Compensation laws in some states honor claims for total deafness but not partial loss of hearing. Only a partial loss is the usual result of excessive occupational noise exposure. Workman's compensation provisions in other states cover partial loss of hearing due to noise but require the claimant to be six months away from the job before settlement can take place. A comparison of state laws covering industrial hearing loss presents a confusing picture which is best illustrated by differences in the amount of benefits received for such disablements. For example, Michigan grants a maximum award of $28,500 for total loss of hearing in both ears while Nebraska awards only $3,700 for the same case.

Other estimates on the cost of industrial noise in the United States run much higher than the figures noted by the Committee. U.S. Senator Mark Hatfield, addressing a 1969 noise abatement conference in Washington, estimated the cost at $12 million per day. "The total cost to industry in the U.S.," the senator declared, "has been estimated—in compensation payments, loss of production, decreased efficiency—at more than four *billion* per year."

. . . Legal Compensation

The confusion which exists in the lack of uniform noise standards, uniform criteria establishing how long an average person must be exposed to high-level sound before he has suffered a permanent hearing loss, and finally how much loss is compensable as personal injury, finds its roots in the workman's compensation codes. These were formulated originally to handle cases of clear-cut injury—such as loss of a hand, a leg, or an eye. Obviously, hearing loss from noise comes upon a person gradually and may result from such a wide spectrum of causes that establishment of specific legal damages is difficult.

"Although the medical history of noise-induced hearing loss is fairly clear-cut and there is no doubt in the minds of anyone that hearing loss will result from noise exposure," states Dr. Glorig, "legally the problem is much more complex and not as clear-cut. Terms such as *disability* and *occupational disease,* when used in the legal sense have a meaning entirely different from the medical meaning. . . . Within a legal framework, loss of function or injury to the human body becomes a disability only if defined as such by law. What might be classed as an occupational disease or a disability in one state will not necessarily be so classed in another."

Joseph A. Sullivan, legal advisor to Liberty Mutual Insurance Company, comments: "Prior to the passage of compensation laws, if an injured worker sued his employer, he would have to prove that the injury was due to the negligence of the employer in order to recover damages. In addition, the employer had three potent common law defenses: contributory negligence, assumption of risk, and negligence of fellow servants. Many injured workers went uncompensated and the burden of their maintenance, during long weeks of convalescence from injury, fell upon local welfare and charity agencies."

Such a system was bound to result in protracted legal haggling between employer and employee. Consequently a system evolved which allowed the employee and employer to settle these problems on the basis of fixed statutes which became known as workmen's compensation laws. Many of the problems arising from these laws, and their necessity, still have not been settled, but as the Supreme Court of the State of Washington commented in 1916, "Present laws came of a great compromise between employers and employee. Both had suffered under the old system, the employers by heavy judgments of which half was imposing lawyer's booty; the workmen through the old defense of exhaustion in wasteful litigation. Both wanted peace. The master, in exchange for limited liability, was willing to pay on some claims in the future where in the past there had been no liability at all. The servant was willing not only to give up trial by jury, but to accept far less than he had often won in court, provided he was sure to get the small sum without having to fight for it."

One basic weakness in the workmen's compensation laws has been the interpretation that payment was not made for bodily injury or loss of physical function, but to replace loss of earning capacity. The laws provided small specified payments that offered a means of supplementing the workman's reduced earnings. No attempt was made to evaluate the worth of the member that had suffered injury, and the law implied there must be a specific date of injury which produced a loss in wages. In most cases, neither of these conditions exist in a case of occupational hearing loss.

As industrialization continued and more workmen were employed, it became evident that many industrial processes produced specific diseases as a direct result of occupation. Such conditions could not be defined as an injury. They did not have a specific date of injury, nor in many cases did they produce a wage loss.

Most problems encountered in enacting fair compensation laws resulted from the difficulty of finally establishing that occupational disease may be compensable. Noise-induced hearing loss is a good example of this. It is unquestioned that hearing loss reduces a man's capacity to live a normal life, and in some instances may reduce his earning capacity.

As time passed and the question of workman's compensation became more important, the coverage of the statutes gradually broadened. A milestone case occurred in 1948 when the state of New York declared that noise-induced hearing loss was an occupational disease and deserved compensation. A similar case went through the courts in the state of Wisconsin in 1951, and the attention focused upon the hearing loss question resulted in a new law written expressly for noise-induced hearing loss. It became effective in July 1955 and Dr. Glorig terms it "the most advanced law in the country today. As professionals," he told members of his own profession, "we must urge that the industrial community do everything possible to preserve human functions. We should be impartial in our decisions regarding all members of the industrial community, which includes employee and employer alike. Our concern is conservation, not compensation."

Another commentary on compensation for hearing loss comes from attorney Edmond D. Leonard of Orinda, California, specialist in industrial labor law. "Despite all precautions," he said, "some industrial loss of hearing must be anticipated. Just as you cannot unspill milk, so you cannot recover hearing loss to acoustic trauma. The problem legally is one of compensation for an irretrievable loss. Laws, to preserve and enhance man's enjoyment of the values of life, must be delicately but firmly interwoven into the fabric of his environment."

.. Industrial Noise Regulation

On the simple face of the matter, it would appear that American industry long since would have established a reasonable noise environment for its millions of workers, if for no other reason than the estimate that noise is costing $4 billion a year. The fact that this has not happened is due to the same confusing factors which becloud the issue of noise-induced hearing loss at all levels. On the one hand, the medical profession has declined to set standards, claiming this is not a doctor's function; then complaints flood in when standards are set by a nonprofessional group which is not qualified to judge the damage levels. Dr. Alexander Cohen, who heads the National Noise Study of the Environmental Control Administration of the Health, Education and Welfare Department, commented upon this point in 1969: "Despite numerous proposals and endless deliberations, we still lack national standards for controlling noise. Everyone recognizes that excessive noise exposure causes hearing loss, but we don't have a national standard for conserving hearing in noise. As has been the case for many years now, we have differences in philosophies in design rules or standards, we have differences in schemes for measurement, and of course, we lack sufficient data to obtain consensus in establishing standards. Now, owing to these hangups in standards development, it is not surprising that regulatory bodies charged with responsibility for protecting the health, the safety, the well-being of people, set noise limits and other requirements that to the more knowledgeable seem unrealistic, ineffective and nonenforceable."

Dr. Cohen also called attention to the problem that many small U.S. companies would be unable to afford the cost of a genuine noise abatement program. About 80 percent of the American work force is employed in small companies with less than 500 people on their payroll. He called upon industry in general to join with government to supply the additional work needed "to plug the gaps in criteria formulations, to resolve outstanding issues and measuring procedures, and generally increased knowledge in the subject of industrial noise.

"Industry itself," Dr. Cohen said, "can provide excellent sources for obtaining such data, either collecting it themselves or allowing others an opportunity to do it for them. Will industry open its files for such purposes? Will it allow such investigations to take place in its facilities, or will it, due to the threat of an avalanche of compensation claims for noise-induced hearing loss or other noise-related problems, remain aloof?"

Some states, including California in 1962, have established some noise limit criteria for industry, but such criteria differ widely from state to state. It is peculiar, but also perhaps normal for the democratic process, that the first approach to federal regulation of noise limits has been made under a law which is only partially and indirectly enforceable, and then only after the law had been in effect for more than 30 years.

This is the Walsh-Healey Public Contracts Act passed by Congress in 1936. The law contains safety and health regulations which must be met by private contractors holding government contracts of $10,000 or more. Under these regulations the federal government through the Secretary of Labor can require industries to abate occupational health hazards. Unfortunately, until 1968 the rules made only vague reference to noise problems. In November that year, following recommendations by the Committee on Environmental Quality, then Secretary of Labor Willard Wirtz issued a proposal to limit industrial noise to a flat 85 dB and opened the question to hearings. (The proposal did have some leeway in it allowing noise levels up to 92 dB until January 1, 1971.) Of 115 statements filed with the Department, 85 percent opposed the 85 dB limit as being unreasonably quiet. Most complaints came from representatives of mining companies (including sand, gravel, and lime), steel companies, machine tool manufacturers, textile companies, and paper and forest product companies.

It was not until May 20, 1969 that Secretary of Labor George P. Shultz effectuated a Walsh-Healey regulation limiting industrial plant noise—33 years after original passage of the law. The new rules, rather than setting a flat limit on in-plant noise, scale the permissible levels according to the number of hours of exposure per day. Although the sound level is higher than many investigators recommend, the rules permit exposure of 90 dB through an eight-hour day. The rules range upward to permit 100 dB for two hours a day, and the highest level of 115 dB if the exposure is 15 minutes or less per day. The regulation also specifies that individual exposure to impulsive or impact noise should not exceed 140 dB.

THE INDUSTRIAL DIN

"When employees are subjected to sound levels exceeding those listed above," Secretary Shultz stated, "feasible administrative or engineering controls shall be utilized. If such controls fail to reduce sound levels within the above levels, personal protective equipment shall be provided and used to reduce sound levels to those listed." The penalty against government contractors which violate the rules and fail to make correction is loss of government contracts for a period of three years. While these new rules do not govern the level of sound in all American industries, their strict application to government contractors is certain to cause the remainder of industry to take notice.

The new Walsh-Healey regulation is a partial answer to the questions of overall loudness versus time per day of exposure, even though the permissible levels appear to be a high compromise. The rules, however, do not appear to consider the specific frequencies where noise seems to do the most damage, nor do they tackle the question of how many years exposure may be endured without harm. Since noise varies so widely from one situation to another, and individual hearing acuity is even more variable, the latter two factors may never be codified as rules. This is particularly true because as more and more powerful machinery is put to work in industry, the spectrum of noise continues to change, probably toward the upper frequencies. As for years of exposure, it is difficult enough to measure the past without trying to anticipate the future, because not enough people have remained in single industries and single occupations in single locations long enough for statistical evidence to become convincing. As examples, twenty-five years ago very few people were working in factories manufacturing either rocket motors or television sets.

On the other hand, some industries have changed so gradually through the years that some valid surveys have been taken. One instance, cited both by the U.S. Public Health Service and the British investigator, William Burns, is that of the jute-weaving industry. Workers in the survey ranged from one year to more than 50 years of exposure to sound which ranges from 82 dB in the very low frequencies, up to 93 to 98 dB in the 500- to 6000-Hz range, and tapering back to 80 dB at 16,000 Hz. The peak intensity of noise occurs between 1000 and 2000 Hz.

Subjects tested at the end of their first year of exposure already showed a beginning permanent hearing loss in the 4000-Hz range, and throughout the years covered by the tests, the hearing loss continued to be deepest in this frequency range. Subjects who had been exposed to the weaving machinery noise for five to nine years showed a 30-dB threshold shift in the 4000-Hz range, and this hearing loss deepened to 50 dB among persons who had worked in the mills from 40 to 52 years. At the same time, the survey showed that the rate of the hearing loss in the 4000-Hz range began to slow after about 15 years, but deepened at a faster pace thereafter in the frequency ranges between 500 and 3000 Hz. Although hearing loss patterns in other industries would not necessarily conform to this survey result, estimates have been made which indicate that high-level sound would cause a 25-dB loss at 1000 Hz after 45 years exposure, with 45 dB at 2000 Hz, and 50 dB in each of the 3000- and 4000-Hz ranges.

As pointed out by the Public Health Service in 1967, workers usually are exposed only to intermittent noise fluctuating both in intensity and pitch. Under these conditions, it is difficult to measure the amount of noise suffered so that some judgment can be made about the harmfulness of the exposure. Resembling radiation badges, noise dosimeters now are being developed which can be worn by an individual for purposes of measuring the total amount of sound energy to which he is exposed. These estimates of intermittent noise exposure, when correlated with hearing loss, will provide a basis for establishing more realistic noise limits.

While noise limits and standards remain in a fairly ill-defined limbo, much of American industry already has taken steps to remove the grossest noise from the worker's environment on the simple theory that a pound of hearing conservation is worth a ton of hearing loss claims and lawyer fees. Most industrial managers also are sincerely concerned about the personal welfare of their employees, over and above the legal obligation which may be involved. The Dow Chemical Company, as one example, has had a hearing conservation program in effect for fifteen years. "We feel the work environment is the larger part of an overall engineering system to make a profit for Dow,"

THE INDUSTRIAL DIN

commented Edward Schneider, acoustic engineer with the company, in a panel discussion of industrial noise conducted in 1969 by the National Council on Noise Abatement. "We have been practicing hearing conservation since 1954, and we've done it, not because of compensation costs, but to reduce environmental hazards to health." Men assigned to noisy jobs are watched carefully, given personal protection, and replaced periodically. The engineer said the company runs audiograms on 2000 to 3000 employees per year. In regard to noise control, as distinguished from hearing conservation, Dow uses noise control specifications in the purchasing of all equipment suspected of being noisy. This is becoming common practice in many industries.

Ideally, noise control will be achieved in industry when buildings are designed initially with low sound levels in mind, and *additionally* designed with noise-controlled machinery placed in proper locations. It is safe to say that most modern industrial plants are built in this way, but for the thousands of factories that are ten years old or older, noise control must be, for the most part, a cut-and-try art combining several different techniques. Basically, noise may be controlled by (1) proper design in the noise source; (2) isolation of vibration; (3) use of enclosures or barriers to separate a noisy machine from neighboring workers; (4) use of mufflers or sound-absorption devices in open passages, such as air conditioning ducts; (5) suitable location of noise sources in the surroundings, and (6) the proper location of sound-absorbent material around noisy machines. As a last resort, and the oldest method of all, workers may be provided with earplugs or earmuffs, but historically this hearing conservation technique is unsatisfactory because these devices are uncomfortable to wear and muffle voice communications as well as harmful noises.

The acoustical engineer has at his disposal a multitude of techniques and devices which can lower the level of sound to which industrial workers are exposed. It is equally obvious that better factory and machine design, taking noise into account from the inception of plans, is most satisfactory of all even though this may well place extra expense upon industrialists. The actual cost of controlling noise hazards is difficult to compute, but in one company an elementary program consisting of initial and periodic audiometric examinations plus personal ear protection was reported to cost about $2 per employee. If this were applied to the minimum of 6 million industrial workers who the Committee on Environmental Quality estimates are endangered by industrial noise, the cost would amount to about $12 million yearly. That, however, covers only the cost of ear examinations and buying earplugs.

"Improving noise control techniques is a more positive remedy," the Committee states, "but is also more expensive. Estimated costs for engineering noise control in one industry averaged $26 per decibel reduction per employee. That is, reducing the noise level by 10 decibels in a work area of 100 people would cost $26,000. Actually the cost of engineering noise control could be decreased significantly if such provisions were made in the early planning stages of plant layout or in the design of industry machinery. Needed modifications at this point would not be expected to exceed 5 percent of the total development cost."

Volumes of research statistics now are pouring out of various government agencies, as well as from industry itself. The Public Health Service offers courses on industrial noise and its control periodically. While the amorphous problem of noise control struggles for focus throughout industry, machinery and factories *are* being designed better, and a major improvement seems to be developing in the area of automated machines. The point here is that more and more industrial operations are being conducted by automatic control, removing men and women from the machine environment itself. Still lurking in the wings of that drama, however, is the question of what the displaced persons will do for a living after ears are removed from the trauma of noise.

CHAPTER 10

*In the more civilized city of tomorrow
noisy traffic may be relegated
underground while humans enjoy
a more gracious existence aboveground.*

Toward
Tranquility . . .

IT WOULD SEEM, now that men have walked on the moon and soon will learn to live and work there, that the marvels of American scientific and engineering ingenuity soon could find ways to return quiet and tranquility to our environment. The moon, now that it is part of our domain, is possibly the only part of our environment that promises to remain completely quiet. The first reason is that it will be some years before men really begin making noise there, but second is the fact that the surface of the moon has no atmosphere and sound cannot be transmitted in a vacuum.

The point of this illustration is simply that the only way in which the world could achieve complete quiet would be to surround ourselves, or our machines, with a layer of vacuum. That, of course, is impossible and demonstrates the consensus among scientists, audiologists, engineers, and others that there are no miraculous shortcuts to an overnight cure for the pollution which has been growing insidiously within our environment for many decades. Noise, in many ways, falls within the category of air and water pollution. All are gradual degradations of the world in which we live, and they will not be improved until the value of environment is believed by everyone to be worth more than the tools we have demanded for our material growth and movement.

Dr. Robert Newman of Bolt, Beranek & Newman complains, "We keep hearing, 'Why don't we use an air curtain to stop the noise?' It won't work, but it's a good gimmick. We hear, 'Why don't we throw up an ultrasonic screen around this source of noise?' All of these wishful thinking devices and gadgets and suggestions are in hopes that the laws of Sir Isaac Newton will suddenly be repealed and that we won't have to mind ordinary old physics anymore!"

Although no one can discount the possibility of a happy discovery in the future—such as an electronic technique to cancel vibrations before they leave metal or other material—the way to a quieter world undoubtedly will be found in countless small victories over a long period of time. Needless to say, the search must begin now.

. . . Making Technology the Scapegoat

In recent years it has become a growing fad to blame all our woes from noise to war upon technology. This amorphous word has been made the scapegoat by humanists who say we have the wrong technology, or too much technology, or suggest that technology has outraced the social growth of the human being. Senator Mark Hatfield of Oregon made this point when he recently commented, "Our modern day technology has failed miserably to solve the human problems of our society, and I think there is an overreliance upon technology. Somehow there are those who feel that it will automatically solve all human problems, just like there are those in our country today who think government can solve all of the human problems. . . . We need our technology, not for technology's sake, but using our technology to solve the human problems, using our affluence not for the sake of prestige and status as a material possession and gain, but using our affluence and our economic wealth to build for a more humane society."

W. H. Ferry, the conservationist, also managed to set technology apart from the society which created it when he said, "Unwanted sound is only one of the many aspects of a galloping technology that threatens every part of civilized life. We shall cope with noise successfully when we teach ourselves to direct technology to the fulfillment of man's nature.

"Technology today is in a half-developed and primitive state, so that it detracts as much or more from man's welfare as it adds. Technology is at present a law unto itself, achieving its authority in a mistaken mystique of progress. Technology and its by-products, noise prominent among them, have—I hope temporarily—methodically eroded values that are natural to man—his sense of self-worth, neighborliness, ease, privacy and quiet. The worship of material things, grasping after profit, self-seeking on a scale without parallel in history—all are functions of technology."

Mr. Ferry is correct and makes his most valuable point in stating that technology "is in a half-developed and primitive state." He is totally wrong when he states that technology "is a law unto itself." He is wrong when he says "worship of material things, grasping after profit, self-seeking on a scale without parallel in history" are *functions* of technology.

Those functions are the functions of people. Technology is nothing more than the tool which man has invented to satisfy his obsessive lust for security, affluence, and power. An airplane, a jackhammer, an automobile—all of these inanimate objects, no matter how complex, are completely silent until they are set in motion by people. The supersonic military bomber is only today's extension of the caveman's club. Automobiles, trucks, and trains are nothing more than tools to speed the motion of a man's feet. The chain saw, bulldozer, and pneumatic drill are merely improvements on the axe, shovel, and handsaw. The products of technology are tools which men use for good or evil. If the tools are noisy, it is man's job to change them.

As Mr. Ferry implied, it is the *maturing* of technology that we seek, and technology will mature only as rapidly as the honesty and morality of men. Those attributes require courage, not a hypocritical search for scapegoats and panaceas. Homes will become quieter when the building trades and zoning boards accept new materials and techniques and building contractors are convinced they must spend however much is necessary to return solidity and quality to structures. Automobiles will be quieter when we stop worshipping noise as a symbol of power and politicians learn to resist the enormous pressures of the oil, automotive, road construction, and insurance lobbies. Airplanes will become quieter when meaningful regulation is applied based upon the will of the people, and realistic costs are assessed against the profit (and noise) makers if the regulations are not met. Industry will become quieter when industrialists must pay for the physical and psychological damage created by industrial noise. If these four statements are totally unrealistic, the cynicism which assumes the public will not fight city hall and eventually will accept the deteriorating environment with a sigh of resignation is all too well founded.

. . . The Power of the Public

Advancing technology will solve its own noise problems if the public makes it clear that the cost of quiet must and will be paid. As stated by the Committee on Environmental Quality, "Our technology has reached the stage where, with few but important exceptions we can cope with almost any indoor or outdoor noise problem provided that we are willing to go to sufficient lengths to do it.

"Increasing severity of the noise problem in our environment," the Committee stated, "has reached a level of national importance and public concern. The problem is broad in scope; it affects almost every facet of daily living and not only has broad socio-economic implications but also affects the health and well-being of our citizens.

"Immediate and serious attention must be given to the control of this mushrooming problem, since the overall loudness of environmental noise is doubling every 10 years in pace with our social and industrial progress. If the noise problem is allowed to go unchecked, the cost of alleviating it in future years will be insurmountable."

Recognizing the sanctity of profit margins and the inertia of politicians caught between powerful opposing forces, the greatest hope for the future is exemplified by the increased public awareness—and action—regarding the world of noise. It appears citizen action may prevent the supersonic transport from flying over populated areas until research solves the sonic boom. Apartment dwellers are fighting the flimsy partitions built between them and their neighbors. The public press and a growing number of civic organizations are making a louder cry against noise. Even more important than these scattered rebellions against noise is the tremendous wave of opposition now developing at all levels against environmental pollution in its many ugly forms. For the first time in this century, 1970 may be remembered as the year in which the average man rose up to be counted, demanding improvement of the world in which he lives.

"Despite this public concern," the Federal Council for Science and Technology reported, "the development of noise abatement measures has not received the attention it warrants by governmental authorities or by the scientific and engineering community. Some governmental units have managed to enact noise control ordinances. There is wide variation and poor correlation among such ordinances, which range from being overly restrictive and impractical to completely ineffectual. Many of these ordinances deal with a limited number of common noise violations, but often overlook or completely ignore control of the major noise producers such as aircraft, traffic and rail transportation systems. These ordinances are difficult to enforce because of economic, social or political considerations and the problems associated with detection or proof of violation."

The Council pointed out, wisely, that achieving noise abatement by the filing of private lawsuits is an unsatisfactory procedure because damages great enough to warrant a restraining injunction are difficult to prove, and it is impractical to rely on the initiative of private individuals to abate noise offensive to the general public. The difficulties encountered by an individual seeking to obtain legal relief from noise such as that from freeway, railroad, or aircraft operations were pointed out by James J. Kaufman, jurist, in a paper presented before the National Conference on Noise as a Public Health Hazard in Washington in 1968. In general, he said, the individual in most cases cannot collect damages for mere annoyance because of a body of common law which terms a "legalized nuisance" a general activity which benefits the general public while harming an individual. At the present time the only clear-cut way in which a citizen may collect damages from a governmental agency is to prove that the use of his property has been denied to him. This proof, in the case of aircraft noise for example, is extremely difficult. Kaufman echoed the Federal Council in calling for local, state, and national uniformity in setting standards of permissible noise and then codifying laws to protect individuals when these limits are exceeded.

"There is no doubt," the Council wrote, "that recognition of the noise problem in America has arrived late. With the exception of aircraft noise, the United States is far behind many countries in noise prevention and control. Consequently there is need for focusing serious attention on abatement measures. Our ultimate goal should be the achievement of a desirable environment in which noise levels do not interfere with the health and well-being of man or adversely affect other values which he regards highly.

"The Federal Government must play a major role in achieving this objective. The problem is a public concern, and its alleviation frequently will require actions that transcend political boundaries within the nation. In addition, significant investments will be necessary for research on noise control. The lack of incentive and the understandable reluctance of our industrial and academic institutions to shoulder this financial burden dictates that the Federal Government assume primary responsibility for research relating to the problem of noise and its effects upon the health and well-being of its citizens."

In brief, the basic needs for noise control on a cohesive, national level are research, the setting of uniform standards, and rational laws to protect the citizen against those who insist upon exploiting natural resources and environment without repaying their debt to society. Although those objectives are difficult, far-seeing planners, scientists, and other experts are waiting on the sidelines with technical innovations which can solve many of our noise problems as soon as people and politicians shift their sense of values toward a willingness to pay for quiet. And once the sense of values has shifted, many innovations will be no more expensive than the established way of doing things.

Transportation and communication provide two interlocking examples of what the future may hold. At the moment we are plagued by the noise of power and inefficiency built into our automobiles, trucks, trains, and airplanes. Most of the noise emanating from these vehicles is caused by the requirement for millions of individual power plants which in turn operate by burning carbon fuels. In the case of surface vehicles, the fuel-burning occurs through repeated, inefficient explosions. Additionally, noise results because these vehicles operate on or above the surface, where the sound is most readily radiated to the surrounding environment. We have previously touched briefly upon the electric- or steam-operated automobile as a possible successor to the internal combustion engine. Although development of such vehicles is vigorously delayed by the status quo of the oil and automobile industries, their advent may not be too many years in the future. In 1969, the California State Senate in a surprise move passed a measure which would prohibit the operation of internal combustion vehicles in the state after 1975. Shocked lobbyists succeeded in throttling the bill before it passed the California House, but the shock may also have been convincing enough to stimulate some serious research and development in the direction of quieter (and less smoggy) vehicles.

. . The Metropolis of Tomorrow

Most city planners visualize future communities with all freight, train, and automotive movement underground, with surface areas devoted to pedestrian malls, parks, and quiet places in general. Other city planners visualize the city building which would contain 10,000 to 50,000 people in a single structure with all industry, shopping, schools, and apartments self-contained. The only transportation needed in such a city (except for intercity travel) would be elevators. A man would no longer need to drive 50 freeway miles to and from work. One study, conducted by scientists at the Aerojet-General Corporation, visualizes all in-city transportation powered by electricity provided from a central source. Trains and subsurface vehicles would draw their power from energized guide rails or, looking farther into the future of technology, from power transmitted and received as we send and receive radio waves today.

Although communication equipment normally is not considered part of our noise-making apparatus, future communication techniques may well ease the onslaught of other noise-makers, particularly the supersonic transport. The use of satellite and other advanced communication techniques in conjunction with computer operations promises the future ability for an industrialist to conduct a remote factory operation by long distance, alleviating the need for frequent air visits to bring superintendents and foremen into line. Additionally, two-way person-to-person television will make possible long-distance meetings and conferences, reducing the need for business travel by an important factor.

As another beneficial fallout from the nation's space exploration program, often maligned because of its high cost, we now can look forward to the establishment of manufacturing plants aboard space stations orbiting the earth. Dr. George E. Mueller, of the National Aeronautics and Space Administration, pointed out in 1968 that weightlessness in orbital flight offers "intriguing" possibilities for new materials, products made more precisely, and new ways to process materials. For example, liquid (including molten metal), when it is floating weightless in space, takes the shape of a perfect sphere. Metal ball bearings could be manufactured in space with precise spherical tolerances impossible on earth, yet at a cost, including transportation, less than we can now achieve. "Perfect bearings," Dr. Mueller said, "would reduce friction and noise levels in our machines to the vanishing point."

These are but examples of what systematic thinking and planning can accomplish toward a world of quiet and tranquility, *if* quiet and tranquility are brought to the fore as overriding values in our society. If a city enforces its noise laws, motorcycles, automobiles, trucks, and jackhammers will be muffled. If the law is loose, or unenforced, the noise will continue. If citizens allow freeways to be built through their communities indiscriminately, they will have noise; if they resist the routing of freeways through quiet areas, eventually they will be heard. If a construction contract contains a no-noise clause, the bulldozers and earth-movers will wear mufflers. Thus, throughout, consistent human control over machines and their use is the key to greater quiet.

The heart of the entire environmental problem—in America and throughout the world—is the tremendous growth in numbers of people, their increasingly rapid concentration in metropolitan centers, and their insatiable appetite for mechanical work which must be done in manufacture, transportation, and freight hauling. During the past century, men moved into the city from the farms seeking education, work, and general enrichment. In the past generation, men have fled from the cities, which grew to be monsters of noise and pollution. In the next generation there will be no room left for escape into quiet and tranquility. Dr. Constantinos Doxiadis, one of the world's outstanding metropolitan planners, states the problem very clearly: "We have failed in human values," he said. "We have failed in building a better operating society. We are heading for disaster. We do not have yet a sign in any city in the world that the situation is under control and getting better. We have many good efforts. We have many good projects. But the situation, in human terms, is getting worse.

"In the growth trend," Doxiadis continued, "we demolish, we choke the center of our cities. Why? Because we allow a city which was prepared to stand up to the pressures of half-a-million people to get 5 million people and later 20 and then 50 million. In cities we try to do everything downtown. We bring all of our machines, all of our pressures, all of our cars downtown, and we choke downtown to death. We are leading our cities to disaster. There is no way out unless we change our course."

Changing course, Doxiadis believes, is possible. Cities can be designed within greater cities, so that "human scale" units of perhaps 50,000 people are linked loosely with other similar units. Doxiadis sees cities of the future built for happiness and safety, where men can re-energize themselves instead of defeating themselves on the nerve-wracking trilogy of tension, noise, and pollution.

Dr. Athelstan Spilhaus, president of Philadelphia's Franklin Institute, believes, with Doxiadis, that preservation of the natural environment requires thinking of pollution in a sense that embraces all the ills of a city, using *disease* as an antonym of *ease*.

"Cities grow unplanned," Spilhaus remarks. "They just spread haphazardly. By planning now, the advantages of high-density living can be preserved without the ugliness, filth, congestion and noise that presently accompany city living. The urban mess is due to unplanned growth—too many sick for the hospitals, too much crime for the police, too many commuters for the transport system, too many fumes for the atmosphere to bear, too many chemicals for the water to carry."

Clearly, if our civilization is to grow, we must set objectives for quiet and tranquility, and pursue those objectives—with all of the tools of humanity and technology—as persistently as we worked to set foot on the moon.

THE NATURE
OF SOUND

CHAPTER 11

*In the young field of noise control,
the acoustic engineer must use
measurements of sound ranging
from the purely psychological
to the purely physical.*

Dimensions
of
Sound ...

SOUND IS SO much a part of our normal daily lives that few of us pause to analyze it. We may think of a sound as uncomfortably loud or inaudibly soft; lovely as in a strain of music or harsh and annoying as in the passage of a low-flying jet plane. Melodious chimes may call us to the pleasant task of eating dinner, or the shriek of tires on pavement may urge us to assist someone injured in an auto accident. During every waking moment of our lives our ears receive and our brain interprets a mixture of thousands of individual sounds which vibrate constantly in complex interwoven ripples through our atmosphere. It is doubtful if anyone—except the totally deaf— has spent a waking hour in complete silence. In fact, a person feels vaguely uneasy in the presence of complete quiet, such as that experienced inside an anechoic chamber which has been scientifically insulated to dampen all sound vibrations and reverberations.

. . . Sound Wave Behavior

Basically, sound originates from a vibrating object such as a bell, a guitar string, or air passing across a speaker's vocal cords. This vibration is imparted to the molecules of air immediately adjacent to the source and then moves outward in all directions in regular waves roughly analogous to ripples which spread after a stone is tossed in a pond.

In physics, a wave is defined as a disturbance initiated at some point and transmitted to other points in a predictable manner determined by the physical properties of the medium existing between the points of observation. Sound waves are a particular form of a general class known as *elastic* or *compressional* waves. Elastic waves can exist only in a medium which has both mass and elasticity. The atmosphere is such a medium. The mass, or inertia, permits a molecule which has been displaced by vibration to impart that motion to another molecule. The elasticity tends to pull the displaced particle back to its original position (as a spring) after the wave has passed.

The speed of sound depends upon two factors, the density of the medium and its elasticity. The more elastic a medium, the greater the speed of sound. The denser a medium, the slower the speed of sound. However, the speed a sound wave travels depends upon a combination of these two factors. Steel is 6000 times denser than air, but it is also two million times more elastic. Therefore, sound travels much more swiftly through steel than through air. Sound travels about 16,400 feet a second in steel, 4700 feet a second in water, and 1100 feet a second (about 750 miles an hour) in sea level air (about 14.7 pounds pressure per square inch) at freezing temperature (32° F.). Because of increased elasticity with temperature increase, the speed of sound increases about one foot per second for each one-degree increase in temperature. Air density decreases with altitude above the earth, but temperature and elasticity also drop sharply so that the speed of sound decreases to about 650 miles an hour at the altitudes where most jet airplanes fly. Sound will not travel at all in a vacuum, such as in outer space, because there are no molecules to bump each other and carry the sound message along. This is why, when Neil Armstrong and Buzz Aldrin walked on the moon in July 1969, the only way they were able to communicate was by radio. No matter how loud they shouted, even if their helmets had been removed, they could not have heard each other on the airless moon. In like manner, there will be no sounds of welding, riveting, hammering, or sawing, when the first space station is constructed in orbit around the earth. Only when the station is provided with an atmosphere will sounds become audible inside.

Two major classifications of sound waves are used in acoustics: *plane* waves and *spherical* waves. The names are derived from the geometric description of the wave front—that area occupied by the forward part of the pressure disturbance. A *plane* wave results, in general, if the source is large, if the measuring point is at a long distance from the source, and if there are no reflecting surfaces between the source and the measuring point. Under these conditions, the sound wave will move across the countryside as though its front were a moving flat wall. The *spherical* wave is transmitted outward in all directions from a central source, just as light and heat radiate outward from the sun. As an example of the difference, a person living near an airport would hear the sound from a jet plane on the ground as a plane wave. The sound from the same jet in flight radiates outward in an expanding sphere. The difference between the two is important because in a spherical wave, the amplitude or intensity of the sound decreases according to the inverse square of the distance from the source. In other words, as the sound waves move outward in all directions, the pressure amplitude decreases because the waves are dispersed over an expanding area. In the case of the plane wave, theoretically the pressure amplitude of the sound does not change with distance unless there is some form of absorption or dissipation in the medium through which it is traveling. This is almost always true to some degree in the atmosphere, and the amplitudes of plane waves decrease with distance and spherical waves lose energy at a greater rate than the inverse square law would indicate.

As a practical rule, sound in the open air always decreases as distance increases from the source, although fluctuations in wind and air temperature and the presence of rain, snow, or fog can make large differences in the loudness and quality of sound heard by different people in different but equidistant positions from the sound source. The strength of sound also decreases when the waves strike any surface. The energy in the moving molecules is dissipated as heat. The total amount of energy in a sound wave is usually quite small, measured in fractions of an electrical watt except for extremely loud noises.

... Pitch and Amplitude

Sound has two primary dimensions, pitch (or frequency) and amplitude, which bears upon the intensity and loudness. The latter two terms are not synonymous as we shall see later.

The pitch or frequency of sound, from high to low, is measured in terms of how rapidly the sound source vibrates. For example, if a tuning fork vibrates 600 times per second, the resulting sound waves have a frequency or pitch of 600 cycles per second. Until recent years, the resulting abbreviation—cps—was the standard designation for frequency. More lately, however, cycles per second have been renamed "Hertz" (abbreviated Hz) in honor of H. R. Hertz, a famed German physicist of the nineteenth century. Hertz opened the way between 1886 and 1888 for the development of radio, television, and radar with his discovery of electromagnetic waves. This Hz designation therefore is used in this volume for all references to sound frequency.

The normal young human ear can hear sounds with frequencies from 20 up to about 20,000 Hz (vibrations per second). Older people or those with impaired hearing often fail to hear the higher frequencies, and there are, at the opposite extreme, rare individuals who can perceive sound up to 24,000 Hz. By way of comparison, the lowest note on a piano has a frequency of 27 Hz, the highest note 4000. Middle C is about 256. Some large pipe organs have tones with frequencies as low as 15 Hz. Most human speech occurs in the frequency bands from 600 up to 4800 Hz. The higher audible frequencies may range from harmonics of the high notes on a violin to the whine of a high-speed turbine engine on a jet aircraft.

Sound waves, of course, are not confined to those which can be detected by the human ear. Very low-frequency sound vibrations, if of strong intensity, sometimes may be felt as a fluttering force on the skin of the body. At the other extreme, dogs can hear sound frequencies above human audibility and silent (to humans) dog whistles are common. Bats navigate by the sound reflection of their high-frequency voices echoing from distant objects or surfaces. Sound below human hearing is known as infrasonic; that above is ultrasonic. Ultrasound may have frequencies as high as 500 million Hz, and the M.I.T. Laboratory at Lexington, Massachusetts has produced experimentally sound waves of 70 billion vibrations per second. Ultrasound is useful in many practical and technical ways, such as in undersea warfare and special cleaning operations.

Quality of sound is a characteristic of musical tones. It distinguishes a tone produced by one musical instrument from a tone of the same pitch and intensity produced by another instrument or a human singing voice. Quality generally results from the blending of many frequencies as parts of the primary source also vibrate in harmonics of seconds, thirds, and fourths above the basic tone. Music generally is considered to be regular vibrations transmitted at regular intervals. Noise, by contrast, may be considered as irregular vibrations transmitted at irregular intervals.

Pitch, or frequency, is the measure of the number of sound waves which pass a specific point in a second. Amplitude is a measure of the depth or height of the sound waves above and below a median line. If a tuning fork is struck lightly, it will vibrate in a very narrow band, the resulting intensity of sound is low, and the sound likewise is of low amplitude. If the tuning fork is struck hard, the vibrations are wider, the sound waves are more intense, and the amplitude is high. No matter how hard the tuning fork is struck, or how intense the sound, the pitch remains the same.

Intensity, resulting from amplitude, is the measure of the amount of energy flowing in the sound waves. People often use the words "intensity" and "loudness" as if they mean the same thing, but they do not. Loudness is a psychophysical measurement of the strength of the sensation received by the eardrum and transmitted to the brain. The same intensity of sound may produce different degrees of loudness for different people; e.g., one person may barely hear a sound that seems clear to another. You may remember a grandfather who could not hear voices clearly audible to others about him. However, for any one person, intensity and loudness depend on four factors: (1) the distance from the source of the sound; (2) the amplitude of vibration; (3) the density of the medium through which the sound travels, and (4) the area of the vibrating object. As an illustration of the last of these four, a hammer blow on a large sheet of steel would produce sound far louder and more intense than striking the hammer against a small bar of the same metal.

The preceding pages offer but a rudimentary view of the complexities inherent in the world of sound. Scientists, engineers, and hearing specialists have spent many decades attempting to measure the millions of variables in sound and its effect upon the human being physically and psychologically. The science of sound is young and alive. Each year experts invent new terminology to provide a "shorthand" by which to communicate more easily. New measurement scales and instrumentation simulate more closely the operation of the human ear, its ability to sort wanted sound from unwanted sound, and the interpretation of these myriad sounds in the human brain. In order to gauge more accurately the rising tide of aural pollution, and then establish standards for noise reduction, we must examine the methods of sound measurement.

Measuring Sound

Standard hearing tests measure a person's ability to hear a sound of single tone or frequency. Each ear normally is subjected to electronic tones ranging successively from 500 to 4000 Hz with the intensity measured to establish the threshold of audibility. In our environment, however, sounds rarely ever reach the ear in a pure single tone, pitch, or frequency, but rather in jumbles of frequency, intensity, and loudness. This is basic to the problem of the acoustic engineer who must measure sound before he can establish standards of noise reduction. The problem, indeed, is so complex that specific electronic measuring tools frequently must be invented and built to cope with specific noise control problems.

In general, however, there are two basic scales against which sound, or noise, may be measured. One is sound power level; the other is sound pressure level. Sound power is measured in watts of electricity radiated by a sound source. Sound pressure is the measure, in terms of atmospheric pressure, of sound as it passes a point any given distance from a sound source. Both sound power and sound pressure are translated into decibels, the most common term used in acoustic engineering today, but with widely different reference points and considerably different numerical scales. For purposes of this volume, and for simplicity, *all decibel readings have been given in terms of sound pressure level, unless otherwise noted.*

.. Sound Power

The human ear hears and copes with a tremendous range of sound power, but most noise sources in our environment originate from sources radiating less than a single watt of power. The most common reference point in sound power level in current usage is one-trillionth $\left(\dfrac{1}{1,000,000,000,000}\right)$ of a watt (more easily expressed as 10 to the minus 12th power; i.e., 10^{-12}). This power level represents zero on the power level decibel scale, and a sound radiated at this power will not be audible to the most sensitive ear. A very soft whisper, the lowest audible sound, radiates about one-billionth (10^{-9}) of one watt. Moving upward in sound power, conversational voice radiates about

$\dfrac{1}{100,000}$ of a watt, and sounds five to ten times that powerful include household appliances such as dishwasher, garbage disposal, etc. A blaring radio, or an automobile traveling at 50 mph generates sound power of about one-tenth of a watt. Typical of sounds generating a watt of power are the output of a 75-piece orchestra, pipe organ, or small aircraft engine.

From that level, sound power moves sharply upward when we consider such noisemakers as multilane highway traffic and jet airplanes. A military turbojet engine with afterburner, for example, generates 100,000 watts of power and a large rocket engine, such as the Saturn V used to propel Apollo spacecraft to the moon, may produce 10 million watts.

It should be stressed again that these elementary figures indicate the total acoustic power radiated by a sound source and do *not* express the loudness of a noise perceived at a distance from the source. For that purpose the sound pressure scale is used.

... Sound Pressure

Sound pressure converted to decibels is the scale most commonly used today in discussions of noise problems in our environment. Here again, it is essential to understand the reference point upon which the scale is based. As sound *power* is most commonly measured on a scale with one-trillionth of a watt as the zero point, the most common reference point for the sound *pressure* scale is .0002 of one microbar. The bar is atmospheric pressure at sea level, 14.7 pounds per square inch with temperature of 59° F. A microbar is one millionth of this value, so it is seen our sound pressure reference point is $\dfrac{2}{10,000}$ of a millionth $\left(\dfrac{2}{10,000,000,000}\right)$ of the atmospheric pressure at sea level. This .0002 microbar reference point is used because a pressure fluctuation of this value is near the faintest sound audible to young people with normal ears. At this level, a whisper may be heard by young healthy ears and thus .0002 microbar represents the zero decibel level on the sound pressure scale.

Among the many noise measurement scales developed in recent years is that which formed the basis for a series of special summer programs on noise reduction offered at the Massachusetts Institute of Technology, 1953 through 1960. On this scale, now the standard with .0002 microbar as zero point, a 14-decibel (dB) reading represents a pressure fluctuation of .001 microbar; 34 dB (or 20 points higher on the decibel scale) represents a sound 10 times as intense. At a sound pressure equaling one microbar, the decibel reading is 74. A noise one million times greater than that (and sufficient to burst human eardrums) carries a 194-dB rating.

The important thing to remember about the decibel scale is that it is not directly numerical, but logarithmic. For example, a noise of 120 dB is *not* 20 percent more intense or louder than a 100-dB noise; it is *10 times* as loud. This concept is important to the techniques for noise reduction because what may appear to be only a small numerical reduction in decibel levels may indeed mean reducing the sound level by one-half or more. Expressed another way, and as a rule of thumb, a doubling of sound pressure is an increase of approximately 6 dB. (By the same rule, a 12-dB increase means doubling twice—i.e., a four-fold increase—and 18 dB means doubling three times, an eight-fold increase in pressure.) As an example of the above rule of thumb, let us suppose that sound pressure in a home a few hundred feet from a commercial jet airfield registers 120 dB. If that sound could be reduced 6 decibels (to 114 dB) the resulting sound pressure would be half as great.

For the average reader, there is a subjective approximation that an increase of 10 dB would be judged, on average, to make a sound twice as loud. It is even more meaningful if decibel readings may be compared with familiar sounds.

Normal conversation rates about 60 dB on the sound pressure scale. At 80 dB (noise 10 times louder) it has been found in industry that exposure for eight hours a day over an extended period of time will cause a hearing loss.

A kitchen food blender may produce noise of 90 dB. A kitchen with blender, vent fan, cake mixer and dish washer all running at once may produce noise up to 100 dB. A loud power mower will hit 107 dB, a loud motorcycle, 110 dB, and an amplified

DIMENSIONS OF SOUND

rock and roll band may generate noise up to 120 dB. (It is interesting to note that sound pressure from the musical aggregation is a trillion times greater than a barely audible whisper.)

A pneumatic riveter creates sound pressure of 130 dB, and if you were to stand 100 feet from a jet airliner as it begins its takeoff, you would experience 140 dB. Few people in our society experience sounds more intense than that, although as has been previously mentioned, large rocket engines may produce sound pressure up to 194 dB, which is the theoretical limit. Above that, the space between individual sound waves becomes a vacuum.

The sound pressure scale in terms of decibels will be referred to consistently in succeeding chapters, which also will include discussion of the physical and psychological harm which may be caused by noise of varying loudness and frequency.

Although the decibel scale is the greatest common denominator in considering noise and its reduction, the acoustic engineer will consult a vast number of scales and measuring techniques in solving different noise problems. For example, noise outdoors is different in character from that within an enclosed room. Steady-state noise, such as that generated by a jet plane engine, differs greatly from an impulse noise such as gunfire or the sonic boom. Noise at certain frequencies may be more irritating than at others. (Consider, for example, the difference between the high-frequency sound of an electric saw contrasted with the low frequency of distant thunder.)

. . . Other Terminology

Four major concepts relate to the physical measurements leading to noise control: (1) absorption coefficient, (2) reverberation time, (3) transmission loss, and (4) noise reduction.

The sound absorption coefficient of any material represents the ratio of energy absorbed by the material to the total energy striking the material. If the material is a perfect reflector, such as metal, the absorption coefficient is 0; if the material is a perfect absorber, such as thick rock or glass wool, the coefficient is 1. The absorption of sound is accomplished by conversion of energy to heat when air particles strike the surfaces of the absorbing material. In any material, the more surface exposed

Typical Overall Sound Levels
in Decibels on the Sound Pressure Scale*

	dB	
	140	Threshold of pain
Hydraulic press (3') †	130	
Large pneumatic riveter (4')		
Pneumatic chipper (5')		Boiler shop (maximum level)
Rock and roll band	120	
Overhead jet aircraft, 4-engine (500')		
		Jet engine test control room
	110	
Unmuffled motorcycle		
	100	Construction noise (compressors
Chipping hammer (3')		and hammers) (10')
		Woodworking shop
Annealing furnace (4')		
		Loud power mower
Subway train (20')		
Heavy trucks (20')		Inside subway car
Train whistles (500')		Food blender
	90	
10-hp outboard (50')		Inside sedan in city traffic
Small trucks accelerating (30')		
		Heavy traffic (25' to 50')
	80	Office with tabulating machines
Light trucks in city (20')		
Autos (20')		
	70	
Dishwashers		Average traffic (100')
		Accounting office
Conversational speech (3')		
	60	
	50	Private business office
		Light traffic (100')
		Average residence
	40	
	30	
		Broadcasting studio (music)
	20	
	10	
	0	

* 0.0002 microbar=zero decibel level.
† Measurements in parentheses indicate distance from the sound source.

to the air movement, the greater will be the absorption. As an example, porous draperies will absorb a great deal more sound than will wood paneling in a room. The design objective is to produce a fiber of very small diameter so that within a given volume, more fibers can be packed and more rubbing surface area exposed to the sound waves in the air. The two principal ingredients are (1) exposure of a large amount of surface area to the sound, and (2) the movement of air within the material. In an outdoor environment, to carry the example further, sound may be absorbed by leaves on trees, fog, falling rain, or snow.

The technical definition of reverberation within a room is the time required for any sound to decrease in amplitude by 60 dB if the sound source is turned off and the sound allowed to decrease by the absorption within the room. If all surfaces in a room are hard, more time will be required for the sound to dissipate; hence the reverberation time will be longer than would be the case if the room were lined with absorbent panels. From the standpoint of noise control, the problem usually is not how long a time is required for the sound to diminish, but rather the steady sound pressure in the room while the source, such as a teen-age musical group playing in a ballroom, is operating.

Transmission loss (TL) describes the reduction in sound passing through a wall from one room to another. This can be measured accurately if there is no other noise in the receiving room but since most rooms are not completely silent, the transmission loss is measured in terms of noise reduction (NR). The efficiency of a wall as a sound barrier may be increased by allowing less energy to pass through the wall. In efforts to arrive at a single term for rating the sound-muffling abilities of a wall, acoustic engineers have coined the term sound transmission loss (STL), which relates sound frequency to the transmission loss characteristics of the wall.

. . . Psychoacoustics

No matter how definitive scientists become in measuring sound in all its frequencies and amplitude, one very important question remains: How does the sound affect the human being? A further question also may be asked: "After it has been determined what physical damage is caused by noise, what is the psychological response and damage?" In recent years, considerable effort has been devoted to definition of meaningful terms to quantify the relations between the physical stimulus (sound) and the psychological response. Three such terms, now of common usage, are (1) loudness level, (2) loudness, and (3) noisiness.

Loudness level is a hybrid term which involves a psychological judgment but also involves physical measurement of sound pressure level. In general, using this term and technique, a subject is asked to rate the loudness of a noise against a 1000-Hz pure tone stimulus. The relative sound pressure level, measured in decibels, then is judged to be the loudness level of the noise.

True loudness is a psychological measure which has been scaled by a number of experimenters, particularly S. S. Stevens and E. Zwicker, using a unit known as the sone. A sone is further defined as the loudness of a 1000-Hz signal of 40 dB sound power level, about the level of a quiet voice.

The advent of the commercial jet airliner some years ago touched off investigations to establish the relative acceptability of the noise of jet aircraft compared with the noise of propeller aircraft. This research led to development of the concept of noisiness, with the unit of noisiness as the noy. The concept of noisiness then led to the concept of perceived noise level in decibels (PNdB), which has become a common reference, particularly in efforts to determine noise standards and controls in residential areas around the world's jetports. One important point in psychoacoustics is that there is a significant difference between noisiness and loudness; though the two concepts interrelate, they are not the same. For example, a symphony orchestra playing the *1812 Overture* would be loud, but most listeners would not consider it noisy. A jet plane, equally loud, would be considered noisy by almost every one.

The language of the science and engineering of sound is still in the formative state because noise control is still a young discipline. Other terms will be encountered—such as speech interference level, temporary threshold shift, and permanent threshold shift—but these will be explained in their proper context in later chapters.

CHAPTER 12

*Long-term high-intensity noise
destroys delicate aural mechanisms
and may even impair blood circulation
and contribute to heart disease.*

Physical
Harm . . .

ADMIRAL LORD RODNEY of the British Navy was described as being almost totally deaf for fourteen days following the firing of eighty broadsides from his ship, the H.M.S. *Formidable,* during a battle in the year 1782. It is not recorded if his full hearing returned, or whether some permanent hearing loss continued for the remainder of his life.

During the following century the coal-fed industrial revolution required steam engines both for land-based power such as steam locomotives and for ship propulsion, all operating with boilers of riveted construction. The massive noise produced in industrial factories by riveting hammers fashioning large steel plates into steam boilers left a permanent mark in the terminology of clinical medicine. The term "boiler-maker's ear" refers to deafness common among men who worked in the bustling boiler factories and is still applied to workers suffering hearing loss from the noise of industrial machinery.

These two instances are extreme examples of the manner in which gross noise may harm a human being. During the past century, and the past fifty years in particular, many volumes of research data have been gathered and published in an effort to determine what level of noise is damaging, how that damage is caused, and the conditions under which the harm may be temporary or permanent. As previously shown, the air pressure fluctuation caused by sound waves is so small, relative to other forces upon the human body, that even the loudest noise known (with the possible exception of the hydrogen bomb) does not directly injure parts of the body except the ear. Noise may cause physical *change* other than hearing loss, but normally not direct injury.

"The auditory system," states K. D. Kryter of the Stanford Research Institute, "is so much more sensitive to sound than any other structure or organ in the human body that, at least for airborne sound frequencies from 20 to 20,000 Hz, it appears that the ear will suffer irreparable harm before the non-auditory systems of the body are adversely affected or, for the most part, even stimulated.

"In spite of the great amount of research attention this [noise] problem has been given, there are a number of unanswered specific questions that await answers from painstaking research with humans and animals. Nevertheless the state of knowledge has reached the stage where it can be applied with a considerable degree of confidence in the specification of tolerance exposure conditions for a tremendous variety of noises."

Two reasons why it has taken so long to reach present levels of understanding are (1) the fact that until recently there were not available the electronic instruments for measuring noise in ways that simulated, with fidelity, the capability of the ear to respond to noise, and (2) the fact that the ear, while a tremendously sensitive organ, is at the same time a very complex and tough one.

Because the human ear is both tender and tough—and different among individual human beings in both attributes—it is necessary to examine current knowledge of the ear itself.

. . . The Human Ear

Most readers are familiar with the common cutaway diagram of the marvelous series of physical mechanisms which work together to translate fluctuations in air pressure into nerve impulses which the brain interprets as sound. Probably· the average person associates damage to the ear in terms of punction or rupture of the eardrum. However, hearing loss due to aging or excess noise is not caused by damage to this tympanic membrane. In fact, a rupture of the eardrum usually will heal and may preserve more sensitive parts of the hearing apparatus in the middle and inner ear.

Hearing begins with sound pressure fluctuations in the air which enter the external auditory canal and vibrate the eardrum. Vibration of this membrane moves a lever system of three small bones, the auditory ossicles, in the air-filled cavity of the middle ear. The ossicles are grouped so that the eardrum vibration drives the first, known as the malleus, which drives the second, (the incus); finally, the third ossicle (stapes) sets up vibrations in fluid and certain structures contained in a complex system of coiled canals known as the inner ear. These canals are imbedded within the bone of the skull. They include the semicircular canals, which are concerned with balance, and the cochlea. In the cochlea is situated the end organ of hearing, the organ of Corti, where mechanical vibrations set up nerve impulses in the fibers of the auditory nerve. The information travels to the appropriate parts of the brain and there is perceived as a sensation, probably stored as a memory, and capable of arousing pleasure, annoyance, fear, or other secondary effects.

One of the most interesting facts bearing upon human reaction to intense sound, may be noted here. Two small muscles (tensor tympani and stapedius) are attached to the malleus and the stapes. When intense sound occurs, or when a foreign object touches the external ear canal, these muscles contract, pulling the stapes and eardrum in toward the middle ear cavity. This tightens the tympanic membrane and increases resistance to movement in the chain of bones in the middle ear. Many sounds thus are not transmitted with the full force received, so that the inner ear is partially protected against damage. The action compares roughly with the iris of the eye which regulates the size of the pupil according to intensity of light. Researchers studying this protective aural reflex have found that it is activated most readily by sound in frequencies of 1000 to 2000 Hz, and the sound must reach intensity of about 80 dB above the threshold of the individual's hearing before the reflex operates. When it does, it may reduce the sound reaching the inner ear by as much as 30 dB in some frequencies. The reflex would operate, for example, if someone were to shout or scream near a person's ears.

The fact that a protective reflex exists in the middle ear points to the sensitivity and complexity of the inner ear, which is further protected by the bone of the skull.

The end organ of hearing is contained within a chamber known as the cochlea because of its complex form resembling that of a snail shell. The cochlea is filled with fluid that is set into motion whenever sound waves are transmitted inward by the ossicles which terminate in two tiny muscular windows which close the end of the cochlea. Inside the cochlea is located a central membrane, and upon this membrane is the complex system of cells and tissue known as the organ of Corti. In an oversimplified description, the organ of Corti also incorporates two sets of hair cells, one group numbering about 20,000, the second about 3500. Mechanical stimulation of these hair cells, by waves of vibration caused by sound in the fluid of the inner ear, is the final step by which the mechanical vibration is converted to electrical impulse and thus conveyed through the auditory nerve fibers to the brain.

Although any of the moving parts of the hearing mechanism may be damaged or even destroyed by intense sound—such as an explosion or naval gunfire—it is the hair cells which seem to be most critically affected. It is difficult or impossible to examine this damage in the human ear, but it has been demonstrated by microscopic examination of the ears of experimental animals such as guinea pigs.

Auditory Measurement

Normal hearing is a statistical concept and normalcy in this field has been difficult to determine despite enormous quantities of measurement and correlation over the past half-century. In general, the norm represents the average auditory thresholds, at frequencies ranging from 50 to 16,000 Hz, for normal young people, measured at the entrance to the external ear canal. Because of differences in measuring techniques, instruments, and other factors, it was not until 1964 that the various committees of the International Organization for Standardization were able to establish a set of hearing standards acceptable by all nations. These are averages which provide a mathematical curve against which an individual may be measured. Description of instrumentation suitable for various audiometric tests can be found in such books as *Noise Reduction* by Leo Beranek, et al., published by McGraw-Hill Book Co., Inc., New York, N.Y.

Age and Hearing

Aging apparently is an inevitable feature in the life of the human body. Due to the progressive reduction of the elasticity of the lens of the eye, for example, the minimum distance for distinct vision recedes as age advances. This effect is known as presbyopia. The less obvious change in hearing with age is known as presbycusis, a progressive deterioration of hearing for high tones. At any given age, the increased intensity of sound required for audibility, by comparison with young people, is directly related to frequency. In audiometric terms, the higher the frequency of the test tone, the higher the threshold level of hearing; and the older the person, the greater the hearing loss. Generally speaking, the loss of hearing with age is similar to, but not the same as, hearing loss due to noise. The difference is detectable because noise causes loss of hearing acuity in specific frequency ranges.

As indicated above, hearing loss due to aging occurs more severely at the upper frequencies than in the lower. The hearing loss (or increase in the threshold of audibility) in the average person as he ages from twenty to sixty years amounts to only about 10 dB at the 500-Hz range. At 3000 Hz, the hearing loss at age sixty amounts to 20 dB, and at 12,000 Hz, the loss averages 70 dB. Stated in another fashion, if a healthy young man of twenty years of age can hear a 12,000-Hz signal at 20 dB, by the time he is sixty the tone would need to be 10 to 20 times more intense for him to hear it.

In all groups covering the age range roughly between eighteen and fifty-four years, most researchers have found women's ears to be considerably more sensitive than men's in the 3000, 4000, and 6000-Hz frequencies. In older groups significant differences also have been found at 2000 and 8000 Hz. According to William Burns, professor of physiology at the University of London, the hearing superiority of women over men is not due to a basic difference in the sexes but to the greater noise exposure sustained by male ears in a lifetime of work, military service, or recreation. Professor Burns believes that small-arms fire—in war, hunting, or target practice—is a very important source of noise-induced hearing loss. Some researchers have found that presbycusis, the hearing loss associated with aging, occurs in part at least through deterioration of the middle ear and is therefore distinct from the deterioration due to the effects of noise on the inner ear. In total, adding to the great difficulty of assigning specific values to various hearing loss factors, the effect of aging may be added to the effect of noise upon hair cells of the inner ear, and both added to physical changes occurring in the cochlea and within the brain as a person grows older. As with everything else affecting the individual human being, hearing acuity relates in many ways to the individual's state of health.

Although it is not experimentally possible to assess the weight of various factors in determining the cause of hearing loss with age, one case in medical annals helped serve this purpose by accident. An Amsterdam doctor reported that a man eighty-two years old, seeking help for impaired hearing, showed upon examination to have a normal right eardrum while the left eardrum seemed to be blocked. Probing with forceps, the doctor found the blockage was a wad of impacted cotton. The left auditory canal had been completely obstructed.

When the doctor asked the patient how he happened to be walking around with this ancient plug of cotton in his ear, the man became very angry. He explained that thirty-two years before he had been treated for an illness of the ear and his family doctor had left the cotton in his ear after the last treatment. He complained about this at the time, but the earlier doctor insisted that the cotton had been removed.

After the cotton was finally taken out, thirty-two years later, the Amsterdam doctor found that the unblocked right ear showed a normal pattern of threshold shift due to aging (presbycusis), but the pattern in the left ear showed a better conduction of sound. The forgotten plug of cotton had functioned as an ear-defender against noise for thirty-two years.

.. Hearing Threshold

Anyone who has ridden a farm tractor all day, or listened to the amplified sounds of a rock dance band in an enclosed ballroom for three hours is familiar, after the sustained noise stops, with a ringing in the ears and for a time the inability to hear sounds at normal level. The latter hearing loss is known as temporary threshold shift. If the partial deafness persists, it may become a permanent threshold shift in hearing. The first term is abbreviated TTS; the second, PTS. Still a third term, common in the auditory sciences today, is noise-induced permanent threshold shift (NIPTS). At this point in the study of sound, there is no question that sounds of certain frequency, duration, and loudness will cause TTS. The same sound, if suffered (as in a noisy factory) over a long period of years may also cause PTS, but there is argument within the scientific community as to whether a temporary hearing loss leads by necessity to a permanent threshold shift. One reason why this is so difficult to establish is that it is never practicable to run direct experimental tests with a group of men and women to induce permanent hearing loss by noise exposure. The answers, therefore, are mostly subjective except for correlations which may be reached by using experimental animals. For example, Dr. W. Dixon Ward of the University of Minnesota recently reported upon experiments with twenty chinchillas which were exposed to repeated bursts of moderate-intensity noise and then partially deafened by a two-hour exposure to the same noise at a higher level. No significant correlation was observed between temporary and permanent shifts, but Dr. Ward does indicate that severe temporary threshold shift requires longer time for hearing recovery. His findings also indicate that an intermittent noise causes less threshold shift than a steady one. A noise that is on only half the time may be tolerated for much more than twice the number of hours that could be spent near the noise if it were continuous.

Dr. Ward and other researchers observe that the physical damage associated with temporary and permanent hearing loss due to noise seems to be swollen cells on the basilar membrane within the inner ear, both hair cells and their supporting cells. "Neither the growth nor recovery of TTS is influenced by drugs, medications, time of day, hypnosis, good thoughts or extrasensory perception," Dr. Ward notes wryly. "The locus of the physiological deficit thus seems to be extremely peripheral—at the hair cells themselves."

In a discussion of presbycusis, the normal hearing loss attributed to age, Dr. Ward uses the term "sociocusis" to describe hearing threshold shifts resulting from environmental noise, particularly that *not* associated with a man's occupation. "It was found," he stated, "that even in persons with no recallable history of exposure to high-intensity noise, gunfire, or head blows, the average hearing decreased with age even before age sixty. The hypothesis was therefore advanced that this average loss of hearing represented the toll exacted on a few individuals by the *everyday noises of modern living*." Dr. Ward emphasized this does not mean that everyone suffers hearing loss from the average level of sound. But apparently many people do.

Because of the fact that most information regarding noise-induced hearing loss has been accumulated in relation to industrial noise situations, it is only in recent years that hearing specialists have given serious consideration to sociocusis. For example, if a man were to claim that he suffered hearing decline because of the noise in his workshop, the burden of proof in workmen's compensation cases has generally been upon his employer to demonstrate that this was not so. Since the noisy machine, or machines, is the only *obvious* source of ear-damaging noise, many industrial compensation claims have been granted on circumstantial evidence alone. In other words, as the everyday noise of our society approaches ear-damaging levels, it may be that a man's home environment may contribute as much to his hearing loss as his industrial situation.

Another difficult question has been that of the amount and level of noise necessary to cause hearing loss, complicated by the fact that some people's ears are more sensitive than others. The answer, therefore, may be expressed only in averages with a considerable decibel range of uncertainty. Dr. Ward, for example, refers to the analysis of 6385 individual hearing tests which demonstrated rather clearly that noise of 80 dB, weighted so that low frequencies are discounted, causes hearing loss no greater than that within a non-noise-exposed population of the same age and general social and economic status. On the other end of the noise damage range of uncertainty, industrial data from a number of sources indicate that for relatively steady eight-hour daily exposure, considerable noise-induced permanent threshold shift (NIPTS) occurs when the noise level is 95 dB over a number of years. Therefore, it is fairly clear that a steady level of noise below 80 dB will not cause hearing loss, but some threshold shift will occur in a portion of the population if the steady noise state is above 80 dB. This is significant to those sectors of population which today are exposed to noise levels well above 80 dB, particularly from air hammers, loud motorcycles, and jet planes.

Once again, generalizations are dangerous because different frequencies, as well as different intensities, cause different end results. A "white" noise with equal intensity at all frequencies from 30 to 20,000 Hz, for example, would not necessarily cause hearing loss across this spectrum. Dr. Ward reports that the frequencies which show first and most severe NIPTS are those in the vicinity of 4000 Hz (high voice level), with neighboring frequencies affected later. The reason seems to be that the middle ear transmits the frequencies between 1000 and 4000 Hz most

PHYSICAL HARM

efficiently, so more energy reaches the cochlea in this range. Also, a given area of the basilar membrane is affected by frequencies below its characteristic frequency, but not those above; therefore all of the most intense noise elements affect the 4000-Hz receptors.

Although it has been well accepted that exposure to noise above 90 to 95 dB over a period of time will definitely cause temporary threshold shift, there remain a number of questions concerning the ear's ability and speed of recovery if the noise is removed, or if it is exposed to the noise intermittently. Drs. Donald H. Eldredge and James D. Miller of the Central Institute for the Deaf in St. Louis report there is evidence that it may require about five hours for a 10-20-dB threshold shift to return to normal, but up to sixteen hours to recover from larger TTS. They point out, in relation to the sixteen-hour figure, that if a person's ears still suffer some threshold shift at the beginning of a new day's exposure to the level of noise, the result would be hazardous in relation to permanent threshold shift. The two investigators agree with Dr. Ward's finding that long-term exposure to 95-dB noise would impair the hearing of about 50 percent of the persons exposed and stress that at 105 dB, a steady continuous exposure produces hearing loss in nearly all men who are habitually exposed.

Dr. Ward believes that a two-week period of time is necessary to determine how much hearing loss will be recovered, but little further recovery occurs after a month. Occasionally, when the hearing is damaged by a single experience, such as an exploding firecracker, some slight additional recovery might occur in the second month. There is no indication that a noise-damaged ear will continue to get worse after the noise is removed, and the same is true, according to a number of researchers, concerning the question of increased susceptibility to such change. Conversely, there is no indication that the ear will grow "tougher" or more resistant to noise by habitual noise exposure. Dr. Ward cites research showing that a group of 15- to 18-year-old boys showed more auditory fatigue after working in noise for several months than they did at the beginning of employment.

There appears to be no generally accepted drugs or therapy that will inhibit the progress of permanent noise-induced hearing loss (threshold shift) or cure it. Massive dosage of Vitamin A once was thought to be beneficial but no such action has been verified. Biochemists in a number of countries, especially Japan, are studying the effects of a broad spectrum of drugs and hormonal agents, but so far no clear effect has been demonstrated. Some physicians still recommend a number of alleviative agents and techniques, but as Dr. Ward states, "placebos would doubtless do just as much good.

"Noises above 80 dB are capable of producing some change in auditory threshold," he continued, "and above 100 dB they are almost sure to affect the normal unprotected ear. We cannot reduce noise induced permanent threshold shift once it has occurred by reducing noise exposure, and there is no way to restore normal hearing."

So the evidence is well accepted that excessive noise can cause hearing loss, either temporary or permanent, depending upon the frequency, intensity, and duration. But is it possible for noise to cause other physical damage or harm to the human being? Answers to this question are much less clear and they are complicated by the psychological effects of noise. It is well known that psychological upset can cause physical illness, but in addition to that, it is becoming more and more apparent that loud and continuous noise can have a number of effects, particularly upon the nervous and blood circulation systems.

In 1968 the Committee on Environmental Quality of the Federal Council for Science and Technology reported upon a year-long study of noise and its effects. "Annoyance is not the only piper we must pay," the Committee stated. "There is also a price in human health and efficiency. Prolonged exposure to intense noise produces permanent hearing loss. Increasing numbers of competent investigators believe that such exposure may adversely affect other organic, sensory and physiologic functions of the human body.

"Although by no means conclusive, some evidence indicates that prolonged exposure to intense noise or vibration of ultrasonic frequencies (above the hearing range), as well as at infrasonic frequencies (below the hearing range) also presents a potential threat to health. In short, growing numbers of researchers fear that the dangerous and hazardous effects of intense noise on human health are seriously underestimated."

⇒ Dr. Samuel Rosen, consulting ear surgeon and clinical professor of otolaryngology at Columbia University, is among the list of distinguished researchers who have noted physical side effects of noise, including changes in the circulation system. "It is known," he said, "that loud noises cause effects which the recipient cannot control. The blood vessels constrict, the skin pales, the voluntary and involuntary muscles tense, and adrenalin is suddenly injected into the blood stream, which increases neuromuscular tension, nervousness, irritability and anxiety.

"Rest, relaxation and peaceful sleep are interrupted and often denied those already tense or ill. We now have millions with heart disease, high blood pressure and emotional illness of all sorts who need protection from the additional stress of noise.

"Physiological reactions occur independent of annoyance or any other emotion. During exposure to strong light, the eyes can be protected by closing the lids, but the ears have no lids. They are always vulnerable."

Dr. Rosen pointed out that a meaningless full-spectrum noise of 90 dB caused (among many test subjects) constriction of the small blood vessels, increased blood pressure, and dilation of the pupils of the eye. He stressed that the constriction of the blood vessels occurs each time a noise stimulus is applied at any interval, and occurs whether the subject is awake, or asleep (when the noise itself would not be heard).

Whereas noise, because of the effects noted above, may well be a dangerous irritant to persons already suffering heart disease or atherosclerosis, research suggests the possibility that high-level sound may even be a contributing factor leading toward disease of the heart and blood system. In experiments conducted with rabbits, control animals exposed to 102-dB levels of general noise for ten weeks showed a much higher level of blood cholesterol than animals not exposed to noise. *The diet was identical in both groups.* In addition, the noise-exposed animals developed a greater degree of hardening and clogging of the aorta, the large artery leading from the heart, than the animals not exposed to noise. Cholesterol deposits in the iris of the eyes were more severe and extensive in the noise-exposed than in the non-exposed animals.

The research which will define the exact roles played by noise in the physical health of the human being is still in an early stage. Noise and its effects in a closed and controlled environment, such as an industrial factory and office building, heretofore has demanded most of the attention of acoustical engineers and scientists. However, now that noise *outside* our closed buildings is reaching such high sustained levels, the years of research ahead will unravel the complex interaction of noise with all of the other environmental factors impinging upon the physical human being.

In many ways, it is likely that the psychological complications arising from noise will be found to have a more profound effect upon people than direct physical damage.

CHAPTER 13

*How establish the psychologically
damaging effects of noise
when one man's noise
is another man's music?*

Mental
Friction . . .

A DOG BARKS in the night. Across the street a car door thumps shut. Through an open window throbs the amplified sound of guitar and drums from a teen-ager's radio turned to full volume.

To the owner of the dog, its bark is a comforting signal of protection against intruders. To the owner of the car, the solid thunk of its door is part of the pleasure of owning a new automobile. To the teen-ager, happiness is the state of physical immersion in the stentorian vibrations of modern music.

To a woman trying to sleep all three of these sounds constitute disturbing noise which interrupts her sleep and elevates nervous tension and frustration because she is powerless to stop the noise. It would eventually impair her health if she is well, or intensify her disease if she is ill.

These conflicting points of view on three simple and common sounds are indicative of the tremendous complications which arise during efforts to determine the psychological effects of noise upon individual human beings. The same immeasurable conflict exists with the otherwise well-meaning neighbor who operates his power lawn mower on Sunday morning, or the young man who uses a loud motorcycle to bolster his ego, and indeed, rides it deliberately through a quiet neighborhood for the specific purpose of irritating its residents. The time is virtually past when any of us can move to a quiet place because the noisy machinery of an affluent civilization now invades even the quietest of suburbs. When a noisy dishwasher and the sound of a youngster's radio or hi-fi join forces with street traffic, lawn mowers, jet planes, and finally the sonic boom, the combination at last will become unbearable even if the overall level of sound is not sufficient to cause direct hearing loss.

"Perhaps the citizen lets out an oath when an especially exacerbating noise afflicts him," recently commented W. H. Ferry, former vice-president of the Center for the Study of Democratic Institutions and a leader in the Citizens League Against the Sonic Boom. "He may roar and yelp a bit and declare that there ought to be a law. Then he subsides to a mutter and ultimately to silence, which is precisely what the noisemakers count on. Hence, we have a cacophonic republic.

"We are even more a dinful than a sinful nation, for while we will not tolerate marijuana, the sputtering motorbike on Sunday afternoon and the loudly grinding garbage masticator at 6 A.M. Tuesday are accepted as part of the American way. The noise-makers remind us often that 'one man's music is another man's noise.'

"I do not suggest that the highly technologized community to which we are irrevocably committed can be run in the quiet of a library reading room. I do suggest that we must not base our case solely, or even principally on public health or economic grounds. There is, I fear, some tendency in this direction."

Mr. Ferry, commenting upon the intangible effects of accumulating noise sources, stresses that it is a matter of establishing "what importance noise control and noise abatement have relative to the other values of our urban civilization.

"It is precisely these other values which need emphasis just as much as demonstrations of growing deafness, stress, and inefficiencies arising out of increasing noise. Quiet and privacy are positive values, not to be considered expendable except on showings of the utmost importance. And I cannot be convinced that clatter and racket have in fact supplanted quiet and privacy as values of Americans.

"I am convinced that Americans are putting up with clatter and racket in the mistaken belief that one day everything will settle down and the cacophonic republic will be no more. But this is the unlikeliest of results unless unprecedented steps are taken to restore the positive values of quiet and privacy."

Obviously, every man, woman, or child knows exactly how much noise, or what kind of noise, causes him to be upset emotionally, but how can dependable psychological measurements of danger point be established when, in certain circumstances, noise of a certain level may be desirable and beneficial?

For example, the Rev. John Hughes, pastor of St. John's Roman Catholic Church in Hadley, Massachusetts had a problem because his church was too quiet. His parishioners feared that when they spoke in the confessional their whispers would carry through the stillness to others waiting their turn nearby. Father Hughes called in an engineer from the acoustics consulting firm of Bolt, Beranek & Newman, Inc., and the problem was soon solved. The engineer installed an electronic noise generator in the choir loft, wiring it to a light switch in the confessional. Now, when the switch is flipped, a steady murmur similar to the whir of an air conditioner fills the waiting area outside the confessional, masking the voice inside.

This example is a little-known product of the nation's acoustic industry. Most of its engineers strive to remove sound, but others devote their energies to manufacturing noise. Sometimes these experts fill office spaces with low-frequency background noise—

MENTAL FRICTION

known as "white" noise or "acoustical perfume"—which soaks up distracting conversations and ensures privacy in an office environment. Many noise engineers devote considerable time and talent to insure exactly the right "thunk" of a new car door so that the buyer—who habitually and mistakenly relates this sound to the solid substance in a car body—is convinced he's getting a good buy. Not long ago, the Hoover Company applied plastic material to muffle a whining motor in the company's new portable vacuum cleaner, but they succeeded also in eliminating most of the rushing air noise, which signals to most housewives the dirt-pulling power of a vacuum cleaner. Marketing tests indicated housewives wouldn't buy the machine without the whooshing sound of power, so engineers removed some of the plastic and the whoosh came back. In like fashion, most secretaries equate pride in their typing speed with the sound of the typewriter and one typewriter company found that a silent machine didn't sell well.

So the psychologist must contend with all the foibles of human nature in his attempts to determine the levels and types of noise which do cause annoyance and inefficiency among the majority of people. In general, psychologists have attempted to categorize noise in terms of (1) interference with sleep or rest; (2) annoyance; (3) interference with performance of work, and (4) interference with conversation or any form of communication based on sound.

As Dr. Alexander Cohen of the U.S. Department of Health, Education and Welfare describes it: "Psychological state is a complex of many processes including sensations, perceptions, actions, thoughts, feelings, attitudes, needs and motives. All of these processes find expression in man's behavior and subjective experience. Noise, or unwanted sound, in interacting with these processes may have adverse effects resulting in losses in work performance, sleep disruption, annoyance and irritability."

Dr. Cohen, who heads the HEW's National Center for Urban and Industrial Health in Cincinnati, states the perhaps obvious fact that losses in hearing sensitivity, together with the masking of speech and other desired sounds, is the most significant sensation and perception problem posed by noise.

. Speech Interference

Losses in ability to hear sounds critical to a worker's task or lack of adequate speech communication can degrade efficiency on those jobs having such requirements. Office workers, having varying needs for speech communication in their jobs, tend to rate the severity of noise in nearly direct relation to the amount of voice communication they believe is necessary for their effective performance. The inability to hear warning sounds, signals, or shouts of caution because of noise is a factor in many industrial accidents. As the noise level of our urban environment grows ever higher, it becomes equally obvious that noise may distract a person from attention to emergency auditory signals such as a traffic crossing bell or an ambulance siren. Busy highway traffic is so loud that a motorist often does not hear a fire truck or police car until it is virtually upon him.

Aside from job performance, the masking of speech by noise is a handy measure of annoyance, and one grading scale of this masking, the speech interference level (SIL), has been shown to be a useful gauge of complaints about noise-intrusion in communities. Speech interference level depicts the ability of a noise to mask speech by averaging the sound levels in octave bands centered at 500, 1000, and 2000 Hz. One community study showed that more than 50 percent of the persons interviewed complained of frequent disturbance and annoyance when an SIL value of 60 dB, measured outdoors, was exceeded for at least 80 seconds per hour. Only 15 percent reported annoyance when the same sound level was suffered only 20 seconds per hour. A speech interference level of 60 dB would pose some difficulty in using the telephone. The same noise level, in face to face conversation would require raising the voice in order to have intelligible conversation at a distance of three feet, and using a very loud voice at distances of seven feet.

The penetration of outdoor noises into school buildings and churches has created serious disturbance and annoyance, again owing largely to problems of speech interference. How many clergymen have suffered interruption of their sermon by loud

MENTAL FRICTION

motorcycles on a Sunday morning, and how many teachers have paused in a lecture while a jet plane flies overhead? One school superintendent has reported 40 to 60 interruptions per day in classroom listening at three schools lying within 1½ miles of a major commercial airport. From 10 to 20 minutes per day were lost in each classroom because of this noise intrusion, with a cumulative loss of 12 to 24 total hours per day of instruction time.

... Effect of Noise on Visual Attention

In the perceptual sense, noise affects job performance where close attention—particularly visual attention—is important. Studies of such tasks show job performance improves with reduction in noise level. In a survey cited by Dr. Cohen, cotton weavers, who supervised the operation of automatic machines, had to be constantly alert for breaking threads. Such stoppage required quick detection and repair in order to get the weaving machine back into operation quickly. Reducing noise level through use of earplugs caused an overall production gain of 1 percent, but it was estimated that the improvement in individual efficiency was about 12 percent. (There are as yet no studies to measure the efficiency loss involved if a sonic boom causes a housewife to drop a cake as she is removing it from the oven, or the efficiency improvement in a football team stimulated by the roar of enthusiastic fans.)

Considerable study has been done in England concerning the effect of high-level steady noise upon the *dependability* of a worker. According to Robert Hockey, research psychologist at Cambridge University, experiments indicate noise has an effect only upon certain kinds of visual work, and the effect becomes measurable when the noise exceeds 90 dB. To be adversely affected by noise, the task should be continuous for at least longer than half an hour, it should present task information at unpredictable intervals or in unpredictable places, and it should present information at a high rate. One series of tests, involving the careful reading of dials, showed that noise of 100 dB—such as a drill press or large stamping machine—caused worse performance than a comparatively quiet level of 70 dB, and performance was noticeably worse at the end of an hour's work under the noisy condition. Hockey concludes that high-level noise impairs performance of tasks which require sustained visual attention. However, in some situations, performance is known to improve with the presence of noise. This phenomenon is familiar to industrial psychologists who prescribe music for the auditory environment of workers engaged in monotonous, repetitive tasks. In such instances, the rhythm of music tends to set a regular pattern for production, and psychologists have found that gradually increasing the tempo of music as the day wears along tends to help maintain efficiency when the workers are growing tired.

Dr. Cohen finds, from the correlation of many test results, that other functions affected by noise include the sense of balance and sometimes visual accommodation. One series of tests found that a person's equilibrium, measured while he balanced himself on rails, was impaired by exposure to wide-band noise at levels of 120 dB. The sense of balance was even worse when the man's two ears were subjected to different levels of sound. Another study revealed that high-level noise reduced the speed with which the eye could move through certain angles and focus clearly. It was believed the noise affected the muscles which control the lens of the eye. Although such changes may appear trivial, they become of great importance where men are working under conditions of intense noise and high danger. Such instances are quarry workers using oxygen torches or metal miners operating rock-drilling equipment. In such environments, a momentary loss of balance or impaired vision may mean death or serious injury and, indeed, industrial statistics show that mining and quarrying are among occupations having the highest accident and injury rates per man-hours of work.

Not a great deal of conclusive evidence has shown impairment of action and thought processes due to noise, but it does appear that random bursts of sound, such as that from a riveting gun, are more likely to disrupt performance of any task than steady-state continuous noise, such as a machine which operates with a steady rumble or hum. Noise is more inclined to disturb the quality rather than the quantity of work. That is, it might not

MENTAL FRICTION

alter the total number of responses made or the total work output, but may cause more errors. Some studies suggest that the loss in quality may be due to the subject working faster (and more carelessly) in the hope that a stressful situation can end sooner. For example, astronauts, exposed to 145-dB noise from a jet engine at full thrust, found it difficult to solve simple arithmetic problems and tended to put down any answer in order to end the experiment quickly.

One problem with attempts to measure these relatively intangible effects of noise is that laboratory problems seldom if ever can duplicate the actual working environment. Carrying this point an additional step, it becomes doubly difficult to correlate laboratory findings with the human environment in general. It is enough, perhaps, to conclude that noise in rising levels is a definite irritant to almost everyone and it can cause reduced efficiency in carrying out many types of work.

. . . Attitudes and Feelings

No one knows exactly why chalk screeching on a blackboard causes a chill sensation in the listener, but it is generally accepted that music, through rhythm, tempo, and melody can evoke moods varying from calmness or contentment to states of excitement, elation, or hysteria. Music has been used extensively in psychotherapy either to excite depressed patients or to calm agitated people. Music in industry seems to evoke a positive attitude toward a person's work. Thus music piped into an office building or a department store conveys to office workers or shoppers that their present occupation is a pleasant one. Psychologically, it helps to hide the fact that a clerk's job is dull and monotonous, or that shopping these days often involves hysterical conflict with mobs of people pushing each other about in the search for bargains.

Martial music has been used deliberately in wartime to stir a sense of patriotism in the mass of people and to stir up emotions of conflict and hatred toward an enemy. In the days before World War II, the United States was clearly advised of Adolf Hitler's intentions to conquer the world as the medium of radio carried the clashing, disturbing sounds of martial music and the hysterical voice of Hitler exhorting the Germans to war. In the latter case, it was not at all necessary to understand the German language to suffer fear and apprehension from the sound of Hitler's voice. Then, once we were at war, the sound of music and busy machines in defense factories became the auditory symbols of our fight for freedom.

In a more insidious fashion, this method of building the conditioned reflex with sound has been responsible for a deeply disturbing change in America's youth. Dating from the 1950s, when the advent of Elvis Presley more or less coincided with the transistor radio which a teen-ager could afford to buy and carry glued to his ear, America's youth has been bombarded with the sound of rebellion disguised in words and music designed specifically to annoy parents and sensitive adults everywhere. As children were progressively alienated from proper mores and authority, the music continued to carry them along—like the Pied Piper—until much of today's message in teen-age music is the lure to drug addiction and sexual promiscuity. More than the bad quality of the music itself, it is this information content of the sound which we find so discordant and mentally disturbing.

These examples show how noise can influence attitudes and feeling by virtue of the information conveyed, not by words, but in the quality of the sound itself. Many sounds are annoying not because they are loud but because they convey distress, pain, alarm, or other unpleasant meanings. Doctor-and-nurse conversation in the halls of a hospital, even though conducted in low tones, has been found to irritate patients because the information dealt with other patients' conditions, operations, and symptoms. Also in the hospital, a frequent source of distress and annoyance is the sound of other patients in pain, moaning or calling for a nurse. Such sounds as these constitute one of the reasons why a hospital often is not the ideal place for a patient's recuperation from illness or surgery. Again, the *information* content of a noise such as a fire or police siren causes annoyance because it rouses fear and alarm. This same association of fear with sound may well be the primary basis for most noise complaints around airports. In addition to the sound, residents are fearful that a plane may fall on their homes. Conversely, a person is seldom annoyed (although his neighbors are) by the sound of his own lawnmower, dishwasher, or garbage disposal because these noises indicate the accomplishment of useful work.

The factor of message content in noise thus almost hopelessly complicates the psychologist's task of trying to establish a uniform scale of acoustic annoyance. In other words, one man's noise is another man's music and it becomes virtually impossible to use the statistical evidence from one neighborhood's noise complaints in predicting how other people would react to the same conditions. In general, annoyance grows with increasing intensity and loudness of sound, and is greater for those sounds of higher frequency and pitch. Sounds which occur in random and unpredictable pattern are generally more annoying than steady sounds to which some adaptation may be made. As an example, the early morning crash of a garbage can is enough to infuriate a person trying to sleep in the city, although that same person has become adjusted to the sound of traffic past his home all night long.

Posed, but not clearly answered, is the question of whether mental or nervous disorders might result from, or be aggravated by, noise exposure. Some industrial studies indicate that prolonged high-level noise may cause increased mental stress and maladjustment among workers, with chronic fatigue and neurotic complaints often received from people exposed to daily noise levels over 110 dB caused by such activities as metal cutting or forming. Although it is difficult to establish a cause-and-effect relationship, this does suggest the possibility that better noise control in hospitals and convalescent homes might well contribute to the recovery of the sick. Personality factors, such as introversion, may be associated with poor performance under noise test conditions. On the other side of that equation, a man's pride in his job may far outweigh any discomfort he suffers from the noise environment.

Sleep Disturbance

Everyone needs sleep, the amount varying from person to person and depending upon variable factors such as sleep soundness. Although researchers still do not know why humans need as much sleep as they do, it obviously provides the conditions for restoring body energy and recovery from fatigue. The effects of prolonged sleep deprivation include losses in mental and physical function and efficiency, irritability, hallucinatory tendencies, and idea confusion. The parents of any newborn child could verify this observation, but other than crying babies, sleep-disturbing noise may range from barking dogs to snoring husbands, street traffic, and a sonic boom in the night. Noise, therefore, may cause its greatest damage to modern humanity in the interruption of sleep. Some people are "light sleepers" while others "sleep like a log." Some are aroused by the slightest change in the noise level, or in response to subconsciously awaited sound, such as a heavy-footed teen-ager returning from a late date. Many people are able to fall asleep most easily when a soft broad-spectrum noise, such as that from an electric fan, furnishes a background to mask sharper noises.

A number of hearing studies have shown that greater annoyance results when a person's sleep is disturbed than when he is only talking or listening. This finding, plus the health significance attributed to rest and sleep, has suggested that criteria for noise annoyance should be based upon sleep interruption. However, noise levels required to disturb sleep have been found to vary almost as much as the individual's ability to sleep. K. D. Kryter of the Stanford Research Institute conducted a series of tests to determine what levels of sound could alter the electroencephalogram pattern (a tracing of the electric impulses from the brain) during different stages of sleep. For the deeper stages of sleep, increasing the level by as much as 80 to 90 dB above the awake threshold level often caused no change in the brain wave pattern. Additionally, once a change could be induced by a given sound, Kryter found that several more bursts of sound were needed at higher intensity to awaken the sleeper. For lighter stages of sleep, there was a significant change in the brain wave response when the sound level was only 30 to 40 dB above the subject's hearing threshold. Another researcher presented different levels of 50- to 5000-Hz noise for three-minute periods between 2 A.M. and 7 A.M. to more than 300 sleeping subjects. From 10 to 20 percent of the group were awakened under noise conditions of about 35 dB over threshold, while some were still asleep when the noise level was increased to 70.

Another question still to be resolved is whether noise which is not severe enough to awaken a sleeper still causes an adverse effect indicated by brain wave changes, rapid eye movements, and other physiological measures characteristic of different sleep stages. Individual ability to adapt to noise in the sleep environment is another area of study. The city dweller, who lives with high levels of outdoor and indoor noise, becomes accustomed to it and often sleeps well, but if he goes to the country, a cricket's chirp in the silence may be enough to keep him awake. The story is perhaps apocryphal of the native New Yorker who returned to his boyhood home on Manhattan's Third Avenue after many years of absence. He was unable to sleep in his old bedroom and finally realized it was because the Third Avenue El, where trains had clattered through all his boyhood dreams, had been torn down.

One of the most significant (though predictable) facts emerging from investigation of psychological effects of noise is that irritation is greatest when the individual cannot control the noise. (An aroused sleeper can turn off the alarm clock, but not the neighbor's barking dog.) This factor of controllability is essential to the problem of quieting major noise sources such as street and highway traffic, jet planes, and the sonic boom.

A test of personal reactions to the ability or nonability to control noise was conducted by Dr. David Glass, social psychologist of New York University, and Jerome E. Singer, psychologist at the State University of New York at Stony Brook. They found that random, unpredictable noises caused marked irritation and frustration, as well as dramatic declines in work efficiency, even after the noise was stopped.

"Our study suggests," Dr. Glass commented, "that powerlessness is the cause of the adverse effects. If the noise is predictable and regular, or if the noise is irregular but the subject is told he can shut it off if it becomes too much for him, the frustration and inefficiency do not appear."

The tests involved 50 Hunter College coeds separated into five groups of equal size and subjected to various levels of noise —random and unpredictable. Almost regardless of the loudness of the noise, the girls recorded tension and low ability to solve mathematical puzzles. In a second series of tests, the girls were told (although subjected to identical noise) that they could press a buzzer and turn off the noise if they wished. The second series of tests revealed two to five times the ability to handle frustration as in the first, although few of the girls found it necessary to turn off the noise once they knew they had the power to control it.

The extreme distances at which noise may exert a disturbing influence upon an individual are suggested by a study conducted by J. E. Green and F. Dunn of the University of Illinois Electrical Engineering Department. The two presented findings indicating that very intense weather, such as storms, can affect human behavior, cause accidents, increase the number of suicides, and induce mild malaise and increasing forgetfulness. Aside from the effects of atmospheric pressure change, Green and Dunn believe the behavioral disturbances are caused by very low-frequency sound waves produced by intense weather phenomena occurring some distance away. Infrasonic waves of 1 to 10 Hz are generated by high winds, such as those in tornadoes, and the low-frequency sound waves undergo very little attenuation with distance. The two researchers found a correlation between storms up to 1500 miles away with a higher accident rate and child absenteeism from school in the city of Chicago.

Conclusion

As ONE OF his first actions in 1970, President Richard M. Nixon signed a law passed by Congress creating a Council of Environmental Quality. At the same time, he committed the federal government to an "all-out" fight against further deterioration of our environment.

"We are determined," the President stated, "that the decade of the 70s will be known as the time when this country regained a productive harmony between man and nature. I have become convinced that the 1970s absolutely must be the years when America pays its debt to the past by reclaiming the purity of its air, waters and our living environment. It is literally now or never."

The effectiveness of the new Council, or of President Nixon's commitment to the nation, remains to be seen. Unfortunately, as the power structures of commerce and industry have grown to massive proportions and strength, the average citizen, buried within more than 200 million of his own kind, has grown cynical of political pronouncements. He has watched through a generation while the nation's waters were converted to sewers, the air filled with smog from the burning of carbon fuels, and the land littered with garbage, trash, and millions of rusting automobile carcasses. Despite some fairly ambitious programs here and there in the United States, very little seems to have happened to correct the environmental wasteland which attacks our senses and our very physical welfare. How then can the average citizen hope that a turn-around point is at hand, that the dinful decibels of noise may start decreasing rather than increasing?

The most obvious hope for the future, of course, lies in the fact that finally the small cries of anguish are gaining in number and loudness. The sound of righteous complaint is echoing beyond town hall and the statehouse and finally is being heard in far-away Washington. As we seek a reduction in the nerve-wracking noise of our environment, the only solution is to make ourselves heard so long and loudly that the lawmakers and enforcers cannot ignore *that* noise.

It is easy to make that statement, but obviously more difficult to carry it out. Most of us are so busy with our own affairs, trying to hold our jobs and keep our families together in comfort, that we do not have time to apply constant attention and pressure to city councilmen, state legislators, and congressmen who were elected for the express purpose of protecting our interests. Because most of us do not have the time and energy to watch and pressure the politicians, they are prone to sway and vote in the direction from which they *do* receive pressure. This pressure, in abundance, is applied by special interests in our free enterprise society—interests such as the oil, construction, and automotive industries, which have the logical goal of increasing profits while satisfying the wishes and needs of the public at large. Our task, as a total society, is to impress upon our government and our mechanical servants the fact that environmental improvement and decency must become part of the cost of making a profit.

Many critics of technology complain that our crescendo of unwanted sound—in fact, the entire degradation of our environment—is the fault of scientists, engineers, and industrialists who go blithely along their way inventing and building new machines without regard for what those machines will do to and for humanity. Lord Ritchie-Calder comments that "when the mad professor of fiction blows up his laboratory and himself, that's O.K., but when scientists and decision-makers act out of ignorance and pretend it is knowledge, they are using the biosphere, the living space, as an experimental laboratory. The whole world is put in hazard. And they do it even when they are told not to. . . . We have plenty of scientific knowledge but knowledge is not wisdom: wisdom is knowledge tempered by judgment. At the moment, the scientists, technologists and industrialists are the judge and jury in their own assize. . . . Somehow science and technology must conform to some kind of social responsibility. Together they form the social and economic dynamic of our times."

Other critics go beyond the question of the need for rational control over technical developments and point out that man, after all, is only at the beginning of the Industrial Revolution. We have learned to build machines, but we have not yet learned to build them well.

"Unwanted sound is only one of the many aspects of a galloping technology that threatens every part of civilized life," stated W. H. Ferry, speaking before the 1968 National Conference on Noise as a Public Health Hazard. "We shall cope with noise successfully when we teach ourselves to direct technology to the fulfillment of man's nature.

"Technology today is in a half-developed and primitive state, so that it detracts as much or more from man's welfare as it adds.

Technology is at present a law unto itself, achieving its authority in a mistaken mystique of progress. Technology and its by-products, noise prominent among them, have methodically eroded values that are natural to man—his sense of self-worth, neighborliness, ease, privacy and quiet."

The primary problem is that in our individual pursuit of affluence, freedom, and security, we have ignored the gradual erosion of the natural values. We listened while the background sound level of our environment grew from 20 to 40 to 70 and finally 90 decibels, and then suddenly it was unbearable. It has taken a half-century of intensive industrial and technical development to bring us to this state of affairs. Now we are faced with the urgent need to reverse the trend overnight. That cannot be done, obviously, but it can begin. Technology can cure the environmental pollution it has created. The most elegant and efficient machine is a quiet one.

When we consider the outrages that have been perpetrated upon our living space, let us consider also some of the small things we have accomplished toward alleviating the outrage. By using inadequate technology, an oil company polluted the California beaches and offshore waters with the apparent attitude "the public be damned." Yet the public refused to be "damned" and through a concerted outcry at least received a government commitment that no new offshore oil leases would be granted. The technology which produced the ubiquitous air compressor which offends human ears seemingly at every street corner in every major city also has produced efficient plastic sound-damping materials to help silence the beast. Some Los Angeles residents who sued their city for damages caused by jet airplanes flying overhead have begun to collect monetary damages for their shattered nerves and health. Others may not fare so well in the courts, but the new generation of jet planes are quieter than the old ones and the FAA has finally established regulations which will prevent noise levels from rising higher. The next step, under public pressure, will be increasing stringency of regulation. In the continuing case of the sonic boom, the public voice has been heard so loudly that the supersonic transport probably will not fly over populated areas until technology has solved the problem of the boom.

In the case of the noisemakers, the public welfare will be considered when the manufacturers and operators of all machines become convinced that their profits will suffer if they do not comply with the public demand. A perfect example is the quiet room air conditioner which has been developed because old ones were too noisy.

Tomorrow will be quieter if each of us becomes convinced that we *can* fight city hall. And since city hall or the statehouse or Congress is nothing more than the collective extension of our neighbors, we must enlist our neighbors in the cause as well. If a neighbor's dog barks in the night, it is time to call the authorities and insist, despite innate politeness, that the dog must be kept quiet. If enough neighbors complain about the noise from a power lawn mower on a Sunday morning, eventually the manufacturers will make quiet lawn mowers.

If the flimsy walls of a new house or apartment are sound-transparent, refuse to buy it or rent it. The builders will find a way to do it better. If they complain that the construction trade unions and antiquated building codes prevent them from using new materials and techniques, then it is time to lobby town halls and state legislatures so persistently that the codes will be changed.

If a new freeway threatens your neighborhood, pound the table at the State Highway Department, not just to have the freeway re-routed but to convince the authorities that alternatives to the automobile must be found for the mass movement of people and freight. New techniques are available. Only sloth and political apathy are standing in the way of their utilization.

In 1930, the Council for Preservation of Rural England wrote an invocation for church litany which intoned: "From all destroyers of natural beauty. . . . from all polluters of earth, air and water. . . . from the villainies of the rapacious and the incompetence of the stupid. . . . from all foul smells, noises and sights—good Lord, deliver us!"

The good Lord, indeed, will deliver us from the plague of noise, land pollution, air pollution, and water pollution if we put an active hand to our deliverance. The new national Council of Environmental Quality may prove effective, but it will do so only if we maintain pressure at all governmental levels for improvement in our environment.

The only real danger ahead is a continued apathy through which we might convince ourselves that we can adapt, as we have in the past, to the rising tide of noise. The alternatives to peace and tranquility are madness and death.

References

References for Introduction

1. Henry Still, *Will the Human Race Survive?* (New York: Hawthorn Books, Inc., 1966).
2. Henry Still, *The Dirty Animal* (New York: Hawthorn Books, Inc., 1967), p. 20.
3. Ibid., p. 1 of Introduction.
4. *1967 World Almanac*, p. 379.
5. Lord Ritchie-Calder, "Polluting the Environment," *The Center Magazine* of the Center for the Study of Democratic Institutions 12, no. 3 (May 1969).

References for Introduction (continued)

6. *University of California Clip Sheet,* 28 January 1969.
7. *American Legion Magazine* (August 1968).
8. Ritchie-Calder, "Polluting the Environment."
9. W. H. Ferry, address before National Conference on Noise as a Public Health Hazard, Washington, D.C., June 14, 1968.

References for Chapter 1

1. J. C. Webster, "Effects of Noise on Speech Intelligibility" (Paper presented before National Conference on Noise as a Public Health Hazard, Washington, D.C., June 1968).
2. Sen. Mark Hatfield, address to Conference on Noise, A New Focus for Government and Industry, sponsored by National Council on Noise Abatement, Washington, D.C., February 1969.
3. Theodore R. Kupferman, remarks before conference noted in 2. above.
4. Leo L. Beranek, ed., *Noise Reduction* (New York: McGraw-Hill Book Co., Inc., 1960).
5. Dr. Robert Newman, remarks before conference noted in 2. above.
6. William Burns, *Noise and Man* (London: John Murray, 1968).
7. Public Health Service, *Industrial Noise, a Guide to Its Evaluation and Control* (Washington, D.C.: U.S. Government Printing Office, 1967).

References for Chapter 2

1. "Noise—Sound Without Value" (Report of Committee on Environmental Quality of Federal Council for Science and Technology, September 1968).
2. Dr. Samuel Rosen et al., "Presbycusis Study of a Relatively Noise-Free Population in the Sudan," *Annals of Otology* 71 (1962), 727.
3. Dr. Samuel Rosen et al., "High Frequency Audiometry in Presbycusis," *Archives of Otolaryngology* 79 (January 1964), 18-32.
4. Dr. Samuel Rosen et al., "Vegetative Reactions to Auditory Stimuli," *Transactions of the American Academy of Ophthalmology and Otolaryngology* (May-June 1964).
5. Dr. Samuel Rosen and Olin, "Hearing Loss˜and Coronary Heart Disease," *Archives of Otolaryngology* 82 (September 1965), 236-243.

References for Chapter 3

1. David M. Lipscomb, "High Intensity Sounds in the Recreational Environment," *Clinical Pediatrics* 8, no. 2 (February 1969).
2. *Time,* 9 August 1968, p. 47.
3. *Los Angeles Times,* 4 September 1968.
4. *Wall Street Journal,* 9 April 1969.
5. *Los Angeles Times,* 2 June 1969.

References for Chapter 4

1. *New York Times*, 2 July 1968.
2. *U.S. News & World Report*, 2 June 1969.
3. *New Scientist*, 13 March 1969.
4. Proceedings of Conference on Noise, A New Focus for Government and Industry, sponsored by National Council on Noise Abatement, Washington, D.C., February 1969.
5. "Noise—Sound Without Value" (Report of Committee on Environmental Quality of Federal Council for Science and Technology, September 1968).
6. *New York Times*, 7 September 1968.
7. William Burns, *Noise and Man* (London: John Murray, 1968).
8. Leo L. Beranek, ed., *Noise Reduction* (New York: McGraw-Hill Book Co., Inc., 1960).
9. Walter W. Soroka, "Community Noise Surveys" (Paper presented before National Conference on Noise as a Public Health Hazard, Washington, D.C., June 1968).
10. *Los Angeles Times*, 22 April 1969.
11. *Auto Club News Pictorial* of Auto Club of California (June 1969).
12. *Los Angeles Times*, 8 May 1969.
13. *Time*, 11 April 1969.
14. *Los Angeles Times*, 17 April 1969.
15. *Ventura Star Free Press*, 13 July 1969.

References for Chapter 5

1. *New York Times*, 7 September 1968.
2. "Noise—Sound Without Value" (Report of Committee on Environmental Quality of Federal Council for Science and Technology, September 1968).
3. W. H. Ferry, "To Insure Domestic Tranquility" (Paper presented before National Conference on Noise as a Public Health Hazard, Washington, D.C., June 1968).
4. Walter W. Soroka, "Community Noise Surveys" (Paper presented before conference noted in 3. above).
5. Dorn C. McGrath, Jr., "City Planning and Noise" (Paper presented before conference noted in 3. above).
6. *New York Times*, 18 August 1967.

References for Chapter 6

1. *Los Angeles Times*, 17 July 1969.
2. Henry Still, *Man: The Next 30 Years* (New York: Hawthorn Books, Inc., 1969).
3. *Airport World* (January 1970), p. 16.
4. *Interavia* (February 1969).

References for Chapter 6 (continued)

5. Federal Aviation Agency, *A Citizen's Guide to Aircraft Noise* (Washington, D.C.: U.S. Government Printing Office, 1963).
6. *Interavia Air Letter,* 18 April 1969.
7. *U.S. News & World Report,* 28 July 1969.
8. Alan S. Boyd, speech before Society of Experimental Test Pilots, September 1968.
9. *American Aviation,* 12 May 1969.
10. *Airport World* (January 1970), p. 9.
11. "Land Use Planning Relating to Aircraft Noise" (Technical report of Bolt, Beranek & Newman, Inc., October 1964).
12. *Airport World* (January 1970), p. 9.
13. *Interavia Air Letter,* 2 May 1969.
14. *Flight International,* 25 July 1968.
15. *Aviation Daily,* 7 February 1969.
16. *Interavia* (December 1968).
17. Proceedings of Conference on Noise, A New Focus for Government and Industry, sponsored by National Council on Noise Abatement, Washington, D.C., February 1969.
18. *FAA Aviation News* (December 1969), p. 3.
19. *Airport World* (January 1970), p. 9.
20. Ibid., p. 16.

References for Chapter 7

1. *American Aviation,* 17 February 1969.
2. *Los Angeles Times,* 11 September 1968.
3. *Aviation Daily,* 14 August 1968.
4. *Los Angeles Times,* 9 April 1969.
5. Ibid., 6 December 1968.
6. Ibid., 6 February 1970.
7. Ibid., 10 March 1969.
8. Ibid., 1 January 1969.
9. Ibid., 11 July 1969.
10. *Aviation Daily,* 18 June 1969.
11. "Noise—Sound Without Value" (Report of Committee on Environmental Quality of Federal Council for Science and Technology, September 1968).
12. *Airline Management & Marketing* (December 1968).
13. *Airport World* (January 1970), p. 10.
14. *Interavia Air Letter,* 20 February 1969.
15. *Daily Report for Executives,* 23 January 1969.
16. *Flight International,* 5 June 1969.
17. *American Aviation,* 9 June 1969.
18. *Airport World* (January 1970), p. 15.
19. *Los Angeles Times,* 19 February 1968.

References for Chapter 7 (continued)

20. *Aviation Daily,* 7 August 1968.
21. *Interavia Air Letter,* 11 February 1969.
22. *Aviation Week and Space Technology,* 21 April 1969.
23. *Los Angeles Times,* 27 June 1968.
24. *American Aviation,* 9 June 1969.
25. *Los Angeles Times,* 18 July 1969.
26. "Land Use Planning Relating to Aircraft Noise" (Technical report of Bolt, Beranek & Newman, Inc., October 1964).
27. *Montreal Star,* 12 July 1969.
28. *Air Transport World* (June 1969).
29. *Los Angeles Times,* 22 July 1969.
30. *Aviation Week and Space Technology,* 16 June 1969.
31. Ibid., 24 June 1968.

References for Chapter 8

1. Proceedings of National Conference on Noise as a Public Health Hazard, Washington, D.C., June 1968.
2. Report to the Secretary of the Interior of the Special Study Group on Noise and Sonic Boom in Relation to Man, 4 November 1968.
3. Dr. Garrett Hardin, "Making Error Creative" (Paper, 1968).
4. W. H. Ferry, "To Insure Domestic Tranquility" (Paper presented before National Conference on Noise as a Public Health Hazard, Washington, D.C., June 1968).
5. "Sonic Boom" (Paper issued by the Boeing Co., 1968).
6. K. D. Kryter, "Sonic Boom—Results of Laboratory and Field Studies" (Stanford Research Institute, June 1968).
7. *Time,* 19 January 1970, p. 61.
8. "The SST in Commercial Operation" (The Boeing Co., May 1969).
9. *Los Angeles Herald-Examiner,* 28 June 1968.
10. *Airline Management and Marketing* (September 1968).
11. *Air Progress* (October 1968).
12. *Aviation Week and Space Technology,* 3 February 1969.
13. *Airport World* (January 1970), p. 9.

References for Chapter 9

1. Dr. Aram Glorig, "Industrial Noise and the Worker" (Paper presented before National Conference on Noise as a Public Health Hazard, Washington, D.C., June 1968).
2. "Noise—Sound Without Value" (Report of Committee on Environmental Quality of Federal Council for Science and Technology, September 1968).
3. Edmond D. Leonard, "Control of Industrial Noise Through Regulation and Liability" (Paper presented before conference noted in 1. above).

References for Chapter 9 (continued)

4. U.S. Department of Health, Education and Welfare, *Noise and Hearing* (Washington, D.C.: U.S. Government Printing Office, 1961).
5. Proceedings of Conference on Noise, A New Focus for Government and Industry, sponsored by National Council on Noise Abatement, Washington, D.C., February 1969.
6. Public Health Service, *Industrial Noise, a Guide to Its Evaluation and Control* (Washington, D.C.: U.S. Government Printing Office, 1967).
7. William Burns, *Noise and Man* (London: John Murray, 1968).

References for Chapter 10

1. Proceedings of Conference on Noise, A New Focus for Government and Industry, sponsored by National Council on Noise Abatement, Washington, D.C., February 1969.
2. Proceedings of National Conference on Noise as a Public Health Hazard, Washington, D.C., June 1968.
3. "Noise—Sound Without Value" (Report of Committee on Environmental Quality of Federal Council for Science and Technology, September 1968).
4. James J. Kaufman, "Control of Noise Through Laws and Regulations" (Paper presented before National Conference on Noise as a Public Health Hazard, Washington, D.C., June 1968).
5. *Rotarian* (March 1968).

References for Chapter 11

1. *World Book Encyclopedia*, 1966 ed., vols. 14 and 17.
2. William Burns, *Noise and Man* (London: John Murray, 1968).
3. Leo L. Beranek, ed., *Noise Reduction* (New York: McGraw-Hill Book Co., Inc., 1960).
4. Wayne Rudmose, *Primer on Methods and Scales of Noise Measurement* (Austin, Tex.: Tracor, Inc.).

References for Chapter 12

1. William Burns, *Noise and Man* (London: John Murray, 1968), p. 7.
2. K. D. Kryter, "The Effects of Noise on Man" (Paper presented before National Conference on Noise as a Public Health Hazard, Washington, D.C., June 1968).
3. W. Dixon Ward, "The Effects of Noise on Hearing Thresholds" (Paper presented before conference noted in 2. above).
4. Donald H. Eldredge and James D. Miller, "Acceptable Noise Exposures—Damage Risk Criteria" (Paper delivered in 1968).

References for Chapter 12 (continued)

5. "Noise—Sound Without Value" (Report of Committee on Environmental Quality of Federal Council for Science and Technology, September 1968).

References for Chapter 13

1. W. H. Ferry, "To Insure Domestic Tranquility" (Paper presented before National Conference on Noise as a Public Health Hazard, Washington, D.C., June 1968).
2. Clifford Bragdon, "Noise, A Syndrome of Modern Society," *Scientist and Citizen* (March 1968).
3. Glynn Mapes in *Wall Street Journal*, 10 September 1968.
4. Alexander Cohen, "Noise and Psychological State" (Paper presented before conference noted in 1. above).
5. Robert Hockey, "Noise and Efficiency: The Visual Task," *New Scientist*, 1 May 1969.
6. *New York Times*, 5 September 1968.
7. *Journal of the Acoustical Society of America* 44:1456.

Select Bibliography

Beranek, Leo L., ed. *Noise Reduction*. New York: McGraw-Hill Book Co., Inc., 1960.

Burns, William. *Noise and Man*. London: John Murray, 1968.

Federal Aviation Agency. *A Citizen's Guide to Aircraft Noise*. Washington, D.C.: U.S. Government Printing Office, 1963.

Proceedings of Conference on Noise, A New Focus for Government and Industry. Sponsored by National Council on Noise Abatement, Washington, D.C., February 1969.

Proceedings of National Conference on Noise as a Public Health Hazard. Washington, D.C., June 1968.

Special Study Group on Noise and Sonic Boom in Relation to Man. Report to the Secretary of the Interior, 4 November 1968.

Still, Henry. *The Dirty Animal*. New York: Hawthorn Books, Inc., 1967.

———. *Will the Human Race Survive?* New York: Hawthorn Books, Inc., 1966.

DATE DUE

MAY 28 1971	JY 8 '74	NOV 1 1 2004
NO 9 '71	JY 15 '74	
NO 2 2 '71	FE 2 5 '73	OC 12 '74
DE ~~ON~~	MR 4 '73	OC 31 '74
	M 2 '73	
JA 2 '72		NO 15 '74
FE '72	MY 30 '73	
FE 24 '72	JY 3 '73	MY 22 '75
MR 7 '72	OC 21 '73	MY 26 '75
MY 2 '72	NO 28 '73	MY 21 '76
MY 16 '72	DE 5 '73	OC 13 '76
	2:15	APR 17 '77
JY 17 '72	JA 24 '74	NO 20 '78
AG 18 '72		MY 28 '79
OC 13 '72	FE 7 '74	MY 28 '79
OC 29 '72	FE 21 '74	APR 2 1 1980
	MR 6 '74	
	AP 15 '74	JUN 9 1980
FE 1 '73	MY 14 '74	MAY 2 5 1981